THE EYE OF THE SANDPIPER

THE EYE OF THE SANDPIPER

STORIES FROM THE LIVING WORLD

BRANDON KEIM

Comstock Publishing Associates
A division of
Cornell University Press
Ithaca and London

First published 2017 by Cornell University Press

Printed in the United States of America

Library of Congress Cataloging-in-Publication Data

Names: Keim, Brandon, 1976– author.
Title: The eye of the sandpiper : stories from the living world / Brandon Keim.
Description: Ithaca : Comstock Publishing Associates, a division of Cornell University Press, 2017. | Includes bibliographical references and index.
Identifiers: LCCN 2016056233 (print) | LCCN 2016056974 (ebook) | ISBN 9781501707728 (pbk. : alk. paper) | ISBN 9781501712647 (epub/mobi) | ISBN 9781501712654 (pdf)
Subjects: LCSH: Natural history—Anecdotes.
Classification: LCC QH45.5 .K45 2017 (print) | LCC QH45.5 (ebook) | DDC 508—dc23
LC record available at https://lccn.loc.gov/2016056233

Cornell University Press strives to use environmentally responsible suppliers and materials to the fullest extent possible in the publishing of its books. Such materials include vegetable-based, low-VOC inks and acid-free papers that are recycled, totally chlorine-free, or partly composed of nonwood fibers. For further information, visit our website at www.cornellpress.cornell.edu.

TO MOM AND DAD

. . . by retaining one's childhood love of such things as trees, fishes, butterflies and—to return to my first instance—toads, one makes a peaceful and decent future a little more probable. . . .

—George Orwell

CONTENTS

ACKNOWLEDGMENTS

I am grateful to many more people than this space allows, but I would especially like to thank Betsy Mason, previously my editor at *WIRED*, for encouraging me to follow my curiosities and convictions; Kevin Berger and Amos Zeeberg at *Nautilus*; Brigid Hains at *Aeon*; Kitty Liu at Cornell University Press, without whom this book would not have been possible; and my readers, whose support and appreciation makes it all worthwhile.

Any writer, and especially a journalist, is a conduit for the knowledge of others. Last but most certainly not least, I am indebted to all those people who have so generously shared their expertise and experience.

First publication of each article was as follows: "Organized Chaos Makes the Beauty of a Butterfly," *Nautilus* (June 2013); "Chickadees, Mutations, and the Thermodynamics of Life" as "Evolution's Contrarian Capacity for Creativity," *Nautilus* (July 2014); "The Photosynthetic Salamander" as "The Salamander That Has Photosynthesis Happening Inside It," *Nautilus* (March 2014); "Human Evolution Enters an Exciting New Phase," *WIRED* (November 2012); "'Parallel Universe' of Life Described Far Beneath the Bottom of the Sea," *WIRED* (March 2013); "At the Edge of Invasion, Possible New Rules for Evolution," *WIRED* (March 2011); "A Mud-Loving, Iron-Lunged, Jelly-Eating Ecosystem Savior," *WIRED* (March 2011); "Redeeming the Lamprey" as "The Hated, Invasive

Parasite That's Actually a Key Part of Its Ecosystem," *Nautilus* (April 2015); "Decoding Nature's Soundtrack," *Nautilus* (April 2014); "Being a Sandpiper," *Aeon* (July 2013); "Monogamy Helps Geese Reduce Stress," *WIRED* (November 2011); "What Pigeons Teach Us about Love," *Nautilus* (February 2016); "Chimps and the Zen of Falling Water," *Nautilus* (July 2015); "How City Living Is Reshaping the Brains and Behavior of Urban Animals," *WIRED* (August 2013); "Reconsider the Rat: The New Science of a Reviled Rodent" as "The Intriguing New Science That Could Change Your Mind about Rats," *WIRED* (January 2015); "Monkeys See Selves in Mirror, Open a Barrel of Questions," *WIRED* (September 2010); "The New Anthropomorphism," *Chronicle of Higher Education* (October 2016); "Honeybees Might Have Emotions," *WIRED* (June 2011); "A Day in the Life of NYC's Hospital for Wild Birds," *WIRED* (August 2014); "New Yorkers in Uproar Over Planned Mass-Killing of Swans," *WIRED* (February 2014); "An Eel Swims in the Bronx," *Nautilus* (July 2014); "On Waldman's Pond," unpublished; "The Return of the River" as "Return of the Ghost Fish," *OnEarth* (November 2013); "A Chimp's Day in Court: Inside the Historic Demand for Nonhuman Rights," *WIRED* (December 2013); "Chimpanzee Rights Get a Day in Court," *WIRED* (May 2015); "Medical Experimentation on Chimps Is Nearing an End. But What about Monkeys?," *WIRED* (July 2013); "I, Cockroach," *Aeon* (November 2013); "The Improbable Bee," unpublished; "The Ethics of Urban Beekeeping" as "Forget the Ordinary Honeybee; Look at the Beautiful Bees They're Crowding Out," *Nautilus* (April 2015); "The Wild, Secret Life of New York City," *Nautilus* (September 2014); "Earth Is Not a Garden," *Aeon* (September 2014); "Add a Few Species. Pull Down the Fences. Step Back.," *Conservation Magazine* (October 2014); "Feral Cats vs. Conservation: A Truce" as "The Dingoes Ate My Kitten," *WIRED* (May 2015); "Should Animals Have a Right to Privacy?," *Backchannel* (January 2016); "When Climate Change Blinds Us," *Aeon* (December 2015); "To Bring Back Extinct Species, We'll Need to Change Our Own," *Nautilus* (January 2015); "September 11, Fall Migration, and Occupy Wall Street," unpublished; "Making Sense of 7 Billion People," *WIRED* (October 2011).

THE EYE OF THE SANDPIPER

INTRODUCTION
TREES OF LIFE

What makes us human? It's a favorite question in the recent history of our species, asked by philosophers and scientists, answered in many ways: It's how smart we are. How social we are. Tools, empathy, language, culture. Our DNA. Our hands. The answers are useful less as explanations than as insights into the zeitgeist. Among the essences of humanity to cross my desk lately are irrationality, gossip, and a digital sketch tablet.

Perhaps more insightful than the answers, though, is how we think about the question. It usually contains a subtle, fundamental assumption: that what makes us human is deeply important to us, and also what makes us exceptional. What sets us apart. To be human is to be *different*. My own values and experiences, however, and my work as a science journalist reporting on animal intelligence and evolution, tell me that doesn't make much sense. I suspect that what matters most to me is shared to a significant degree by the average duck.

But to play the game for a moment, there's one answer to that loaded question of *Homo sapiens*' identity that I quite like. It comes from David Abram, a magician and ecologist who locates our uniqueness in a vast capacity for fascination with the world outside ourselves, and especially its nonhuman life. "We really display our uniquely human beauty," writes Abram, when we "allow our attention to move outward, toward the other shapes of sensitivity and sentience with whom we compose this many-voiced biosphere."

To that I would add: As we cast our attention outward, we return with metaphors and similes to make sense of what we learn. Mother Earth and cosmic symphonies and the mysteries of dark matter. Streams of consciousness contain seeds of ideas. Brains compute. These are, scientists say, manifestations of our powerful—and uniquely human, natch—symbolic thought. Our science also describes a tree of life.

There isn't just one. A forest grows from our studies, the earliest limited by reliance on anatomy and preconception: Jean-Baptiste Lamarck's separate trunks for invertebrates and vertebrates, Theodore Nicolas Gill's vertebrates as descended from mollusks, Ernst Haeckel taking liberties with Darwin's famous tree and perching humans at its crown. Later trees benefited from fossil records and then genetics; they're a fine metaphor, conveying evolution's fundamental truth of change across time. Yet they often contain their own conceptual biases: left-to-right or top-to-bottom arrangements, as though change is necessarily progress and species evolve in linear fashion.

These are conventions of graphic form rather than ideology, yet the trees still imply hierarchies of value that are interpretations, not biological reality. My own favorite, drawn by a biologist named David Hillis, isn't even a tree. It's a circle composed of lines radiating outward from the center, something like fungal mycelium or a star coral, except the lines—each an evolutionary lineage, beginning with a single common ancestor several billion years ago and truncating in a ring of modern species—also recall the golden-spiral geometry of a nautilus shell. It's evolution as printed by Apollinaire. Displaying it properly requires a gallery-sized wall, and finding *H. sapiens* in the fine-print ring of species could take hours but for a helpful caption pointing to our name: You are here.

Which isn't to downplay our achievements: antibiotics and rocket ships and supply chains, "Breaking Bad" and Chauvet cave art and Polynesian star navigation, Petra's rock-cut architecture and the Manhattan skyline and the pulsation of 7.5 billion brilliant, emotional, playful beings. And of course the Hillis tree is still a symbol, with its own subtle implications, foremost among them taxonomy's ordinal perspective. It doesn't capture the scent of a violet, much less an ecosystem's energy networks. But encountering it was a powerful moment for me, a lesson in humility and the different levels at which nature is understood.

I first talked to Hillis in 2009 while writing about the possibility of life's origins in ancestors of amoeba-like creatures called placozoans. A few years later, I covered his research on a species of clam that's physically hermaphroditic but genetically male, and reproduces entirely by self-cloning—although, every so often, they steal eggs from other clam species and absorb fresh DNA. The headline: "Crazy Sex Trick Fuels All-Male Clam Species." This is one way to approach nature: a cornucopia of the beautiful and bizarre, of fantastic forms and behaviors that wouldn't be imagined if they didn't already exist.

In my work as a science journalist, I've written relatively few of those stories—not because they're uninteresting but because when I started my career at *WIRED*, nominally a technology publication, they were topically a tough fit. It was easier to focus on biology's dynamics: not tech, certainly, but cutting-edge knowledge. These dynamics—how evolution works, and then how life's systems work—form the first section of this book.

The next section addresses the inner lives of animals, a perspective too often obscured by talk of populations and species. Here I use the term "inner lives" intentionally rather than "animal intelligence" or even "animal cognition," terms that betray a self-serving tendency to qualify experience in terms of mental calculation. Other species might not be as cognitively adept as *Homo sapiens*, at least not on the tests we devise, but they too experience life as individuals, just as we do. Of course there are differences, and these are quite interesting: I'd love to experience time's passage from the perspective of a snapping turtle sleeping through each winter. But the differences are less significant, I think, than what we have in common.

What happens when appreciation of ecology's wonders and animal consciousness collide with 7.5 billion humans in an era dubbed the Anthropocene, the human age, in which our needs and whims have planetary consequences? As do, for that matter, our ethics and habits of thought, the way we understand other lives and balance their needs against our own? Epochal issues, yet realized in our everyday settings: a vacant lot, a dammed river, a pigeon with a broken wing. These stories make up the third and fourth sections of this book.

The stories map my own thoughts and questions over the last decade, during which time I've had the great pleasure and privilege of writing about science. Not all my reporting was about nature; often it involved more immediately utilitarian subjects—public health, genetics, human behavior—and was intended to address injustices and suffering, or simply to help people make better decisions. Yet nature writing was never a fanciful diversion. Always it was intended to enrich readers' lives, just as working on these stories did my own.

To appreciate more deeply a skipper butterfly's flight or a mockingbird's songs, to look at a river and see something that yesterday was invisible, is no small thing. It is a richer experience of being human.

I

DYNAMICS

Like most anyone, I suspect, who pays close attention to the world's news, there are times when my sense of wonder feels tapped out. The endless procession of suffering and hardship and humankind's inhumanity becomes overwhelming; the daily firehose of attention-seeking headlines and sales pitches and social self-curation makes me want to stop paying attention altogether, like a turtle withdrawing into his shell.

Then I'll happen upon some scientific finding—of bacteria living deep beneath the ocean floor, or unappreciated fishes who make their surroundings more verdant—and out pops my head, back into the world.

Of these seemingly esoteric findings, some people might say: What's the use? What do mutation rates in horned toads have to do with anything? They have do with beauty, the joy of discovery, the sheer richness of a planet on which every leaf and blade of grass is fueled by quantum physical processes—and those forests and fields will, if left unattended, reproduce themselves indefinitely, sustaining life's din for as long as our sun shines.

What a world! We are born in a very literal sense onto an enormous spaceship, one whose operation we've just started to understand. And the more we learn, the more extraordinary it is.

ORGANIZED CHAOS MAKES THE BEAUTY OF A BUTTERFLY

Take a look at a butterfly's wing and you can learn a lesson about life. Not that it's beautiful, or fragile, or too easily appreciated only when it's fading—though all that is true, and evident in a wing.

Look very close, at the edge of a pattern, where one color turns to another. The demarcation isn't so abrupt as it seems at arm's length. It's not a line, but rather a gradient.

This is a lesson about uncertainty.

A butterfly's colors come from its scales, each a single cell, pigmented a single hue. At pattern boundaries, scales of different colors intermingle. Transitions and shading are achieved by varying the proportions of the mix. It's beautiful. It is also, in the language of molecular biology, a model for a stochastic mechanism of gene expression.

Each scale's fate is not preordained. Cells on the surface of a swallowtail's wings, for example, were not originally specialized to be yellow or blue or black. Instead they contain genes potentially capable of producing each of those pigments.

What determines the color of each butterfly scale is, in a word, chance. A molecule hits a piece of cellular machinery at just the right moment, in just the right place, and a gene produces a certain pigment. There's no guarantee it will happen. It's a matter of probability and moment-to-moment randomness. (The same probabilistic mechanism underlies the portions of the wings with solid colors, too. In those parts, molecules that trigger genes for one color are present at such high concentrations that the final color outcome is assured.)

In biology, there's a tendency to conceive of randomness as noise, an accidental factor, a product of error—random genetic mutations, for example, are mistakes in a chromosome-duplicating system that's supposed to make perfect copies. Random mutations might be harmful, or insignificant, or beneficial, but they're fundamentally mistakes, a disorderly deviation from an orderly system.

What makes a butterfly's wings so remarkable isn't just that unpredictabilities underlie their colors, but that they've harnessed the probabilities. Randomness and uncertainty are translated into the ordered, functional patterns of a monarch or checkerspot. And in this, the butterfly's wing is not unique, but a manifestation of principles ubiquitous in biology.

Let's put on our *Powers of Ten* goggles and increase our magnification to where cell activity occurs, the level of so-called cellular machinery. We'll have to abandon that metaphor, though: Cells indeed contain complicated, task-performing structures, but the word "machine" is a product of the macroscopic world. We think of machines as rigidly assembled, with predefined purposes. At the cellular level, the analogy falls apart.

Out at the leading edges of theoretical and computational and experimental biology, where known and unknown meet, cellular machines have been redefined. The proteins of which they're made don't fold and unfold and operate according to some stepwise blueprints. Shape and function are exquisitely sensitive to infinitesimal energetic shifts, to the motion of atoms and the forces they exert.

Rather than a cellular factory, then, imagine a restaurant with a kitchen where blenders turn into convection ovens and whisks into knives when someone walks by, raising ambient temperatures by a fractional degree. Imagine that the whole kitchen is like this, that cooks and prep staff, though they move with intent, can't help but wander around—and still the seven-course meals come rolling through the doors.

Undoubtedly this metaphor has its own problems, but it gets the point across: The cellular world is an ever-fluctuating place. It's full of randomness and, when things aren't narrowly random, with uncertainty. Atoms and molecules and gradients change from moment to moment, and proteins with them.

Here one might ask where uncertainty comes from: Is it truly uncertain? Might every molecular fate be predicted, if only we knew the motions and properties of every particle in a cell? Or does quantum physics enter the equation at some level, with all its spooky uncertainties and strange probabilities shaping biology in some fundamental way?

That we don't know, and might never. It's a question too hard to study. Whatever the case, certain molecular activities are, best as we can describe them, random or probabilistic. And we know, up at the cellular level, that a cell's fate—whether an embryonic stem cell becomes specialized for service in kidney or liver, whether a blood stem cell grows up to carry oxygen or identify pathogens—is to some extent stochastic, determined by a signal that may or may not appear.

What's extraordinary, again, is that from all this uncertainty, form arises. Two identical twins, after trillions upon trillions of cell divisions, actually look the same to our eyes. Out of disorder, order—though of course having identical genomes is no guarantee of identical outcomes. Indeed, stochasticity's role in cells became evident when genetically matched yeast colonies, raised in the exact same environments, developed in very different ways.

What seems to explain that, it turns out, is variation in so-called epigenetic responses—processes that alter gene activity according to environment and circumstance, allowing organisms to change their biology in response to life's unpredictable demands. The different yeast colonies had different epigenetics, they responded to uncertainty differently. That might itself have been the product of chance, some *inherited stochastic variation*, though the benefits are obvious. It's evolutionary bet-hedging, a way of increasing the adaptive possibilities for one's descendants, despite their genetic similarities.

Again we see biology using uncertainty, building on it, making it integral to life. And it's evident not only in epigenomes, but in genomes: When we look at our own, which for each of us contain some number of apparently random errors produced by copy-making glitches, we find that the errors are not randomly distributed. Mutations occur at different rates in different parts of the genome.

This isn't the same as saying that certain sequences tend to stick around over evolutionary time because errors there are more likely to cause problems. Instead, the potential for a random error to occur in the first place fluctuates across the genome. At every cellular level, randomness is harnessed.

"Life is a study in contrasts between randomness and determinism," wrote the biochemists Arjun Raj and Alexander van Oudenaarden in the article "Nature, Nurture, or Chance" in *Cell*. "From the chaos of biomolecular interactions to the precise coordination of development, living organisms are able to resolve these two seemingly contradictory aspects of their internal workings."

Do these resolutions occur at even higher levels? Pattern from chance in populations, species, communities, ecologies? In our own lives? We can't look at societies or lives the way we do cells, but certainly we feel it, at some intuitive level. "I think back on the trajectory of my life, and I think: I happened to bump into this person on the train, and it led to this or that," Raj told me. "So many things are unpredictable on a long time scale, though it feels like they are predictable in the moment."

I think on my own life: My parents met on a train. My closest friends came from chance encounters in a subway station, a class, a hockey team, a writing assignment. I can't imagine my life without them, yet each of those meetings was profoundly, unsettlingly improbable. And why stop with friendships? Why not scale up to the level of the universe itself, where order and disorder interpolate in random patterns?

There, perhaps, from the perspective of God or some alien cosmologist or deep time or whatever you use to imagine inconceivable vastness, we might find order yet again. Who knows for certain; we likely never will. But we can look at a butterfly's wing and wonder.

CHICKADEES, MUTATIONS, AND THE THERMODYNAMICS OF LIFE

One of my favorite pastimes while traveling is watching birds. Not rare birds, mind you, but common ones: local variations on universal themes of sparrow and chickadee, crow and mockingbird.

I enjoy them in the way that other people appreciate new food or architecture or customs, and it can be a strange habit to explain. *You're 3,000 miles from home, and less interested in a famous statue than the pigeon on its head?*! Yet there's something powerfully fascinating about how familiar essences take on slightly unfamiliar forms; an insight, even, into the miraculous essence of life, a force capable of resisting the universe's otherwise inevitable tendency to come to rest.

Take, for example, a small songbird known as the willow tit, encountered on a recent trip to Finland and closely related—*Poecile montanus* to *Poecile atricapillus*—to the black-capped chickadee, the official bird of my home state of Maine. To the naked eye, there's not much to distinguish between them. Both are small, with black-and-white heads and gray-black wings, seed-cracking bills, and a gregarious manner. For a long time, they were even thought to be the same species. The only obvious difference, at least with the willow tit I saw, was a duskier olive underbelly coloration.

Which raises a question, asked by Darwin and J. B. S. Haldane and generations of biologists since: Why? Why is a bird, similar in so many ways to another, different in this one? It's a surprisingly tricky question.

Generally speaking, we tend to think of evolution in purposeful terms: There must be a reason for difference, an explanation grounded in the chances of passing on one's supposedly selfish genes. Perhaps those olive feathers provide a better camouflage in Finnish vegetation, or have come to signify virility in that part of the world. As the evolutionary biologists Suzanne Gray and Jeffrey McKinnon describe in a *Trends in Ecology and Evolution* review, differences in color are sometimes favored by natural selection—except, that is, when they're not.

Often differences in color don't have any function at all. They just happen to be. They emerge through what's known as neutral evolution: mutations randomly spreading through populations. At times, this spread, this genetic drift, evenly distributes throughout the entire population, so the whole species changes together. Sometimes, though, the mutations confine themselves to different clusters within a species, like blobs of water cohering on a shower floor.

Given enough time and space, these processes can—at least theoretically, as experiments necessary for conclusive evidence would take millennia to run—generate new species. Such appears to be the case with greenish warblers living around the Tibetan plateau, who during the last ten thousand years have diverged into multiple, noninterbreeding populations, even though there are no geographic barriers separating them or evidence of local adaptations favored by natural selection. The raw material of life simply diversified. One became many because that's just what it does.

Through this lens, evolution is an intrinsically generative force, with diversity proceeding ineluctably from the very existence of mutation. And here one can step back for a moment, go all meta, and ask: Where does mutation itself come from? How did evolution, and evolvability, evolve?

It's a question on the bleeding edge of theoretical biology, and one studied by Joanna Masel at the University of Arizona. Her work suggests that, several billion years ago, when life consisted of self-replicating chemical arrangements, a certain amount of mutation was useful: After all, it made adaptation possible, if merely at the level of organized molecules persisting in gradients of heat and chemistry. There couldn't be too much of it, though. If there were, the very mechanics of replication would break down.

The molecular biologist Irene Chen of the University of California, Santa Barbara, has further illuminated that delicate balance. Her work posits that, as an information storage system, DNA was less error-prone than RNA, its single-stranded molecular forerunner and the key material of the so-called RNA world thought to have preceded life as we now know it.

So, then, one can imagine, early in Earth's history, life evolving again and again, crashing on the rocks of time and circumstance, until finally it hit on just the right mutation rate—one that eons later would produce organisms and species and ecosystems that reproduce themselves and persist across time and chance.

That's the remarkable thing about life: It continues. It keeps going and growing. Barring catastrophic asteroid strikes, or possibly the exponential population growth of a certain bipedal, big-brained hominid, life on Earth maintains complexity, actually increases it, even as the natural tendency of systems is to become simpler. Clocks unwind, suns run down, individual lives end, the universe itself heads toward its own cold, motionless death; such is the Second Law of Thermodynamics, inviolable and inescapable.

Yet so long as Earth's sun shines and genetic mutations arise, evolution may maintain its own thermodynamic law. Black-capped chickadees and willow tits diverge. Life pushes back.

THE PHOTOSYNTHETIC SALAMANDER

Amidst life's profligate swapping and sharing and collaborating, one union stands out: the symbiosis of spotted salamanders and the algae living inside them.

Their uniqueness is no small matter. After all, mutually beneficial relationships between species are legion. Our own genomes are suffused by DNA from other organisms—not inherited from common ancestors, but picked up through the drift of DNA across species. There are cellular mitochondria, the power-generating product of some long-ago meeting, and of course microbiomes, those microbes that account for 90 percent of the cells in animal bodies, and aid in all sorts of physiological processes. Walk through a forest and the trees' roots are intertwined with co-evolved fungi. A symbiotic carpet lies underfoot.

There's no end to it, yet the spotted salamander—*Ambystoma maculatum*, to be exact, common to eastern North America—displays something unique. You can see it for yourself: Should you be so fortunate to find spotted salamander eggs in a vernal pool this spring, take a close and gentle look, and note the tiny flecks of green. That's the algae, *Oophila amblystomatis*. No other land-dwelling vertebrate has an endosymbiote—an organism living inside-out it—that is visible to the naked eye.

One of the first scientists to notice was the biologist Henry Orr, who in the late nineteenth century surmised that the algae did something for embryonic salamanders, though he didn't know what. By the mid-twentieth century, researchers had expanded on Orr's speculations, hypothesizing that algae perhaps fed on cellular waste and in return provided cells with oxygen generated during photosynthesis.

The nature of the symbiosis remained murky, though, and in 2011, researchers found that algae didn't just reside on cellular surfaces, as Orr and others thought. It lived inside the embryos, literally a part of spotted salamander life from beginning to end. They might even be passed from mother to offspring in a transgeneration act of gardening.

Two years later, another finding further illuminated the relationship, almost literally. The researchers incubated salamander eggs in water containing a radioactive carbon isotope; when algae photosynthesized, they produced radioactive glucose. The embryos that developed in that water were also radioactive—but not if they were incubated in darkness, when photosynthesis stopped. Algae's solar-powered activity indeed appeared nourishing.

That hasn't yet been conclusively demonstrated, but there's a substantial weight of evidence. More substantial still is the importance of algae to developing embryos. In their absence, embryos tend to develop slowly and less successfully, and are especially vulnerable to infection. Meanwhile, when outside spotted salamanders, *O. amblystomatis* becomes dormant, forming bunker-like cysts that open up again only when in the presence of their hosts. The species are made for each other.

What might one make of this fact? There are several lessons, one of them immediately practical: Atrazine, a common herbicide, may harm salamanders by killing the algae that live within them. Another, more conceptual, insight might be taken from the hundred-plus years it's taken to reach what we know of this union, which researchers now think might be found in many other amphibians and even in fish. As much as we know, we know very little.

And the clearest lesson of all, as evident as the bright green algal flecks and the yellow dots that give spotted salamanders their name, is this: Nature is not simply "red in tooth and claw," as the old saying goes, a world of vicious competition and ruthless exploitation. It's one of cooperation, of mutual benefits, of the beautiful ways evolution has found to sustain life in alliance.

HUMAN EVOLUTION ENTERS AN EXCITING NEW PHASE

If you could escape the human time scale for a moment, and regard evolution from the perspective of deep time, in which the last ten thousand years are a short chapter in a long saga, you'd say: Things are pretty wild right now.

In the most massive study of genetic variation yet, researchers estimated the age of more than one million variants, or changes to our DNA code, found across human populations. The vast majority proved to be quite young. The chronologies tell a story of evolutionary dynamics in recent human history, a period characterized by both narrow reproductive bottlenecks and sudden, enormous population growth.

The evolutionary dynamics of these features resulted in a flood of new genetic variation, accumulating so fast that natural selection hasn't caught up yet. As a species, we are freshly bursting with the raw material of evolution.

"Most of the mutations that we found arose in the last two hundred generations or so. There hasn't been much time for random change or deterministic change through natural selection," said the geneticist Joshua Akey of the University of Washington, coauthor of the recent *Nature* study. "We have a repository of all this new variation for humanity to use as a substrate. In a way, we're more evolvable now than at any time in our history."

Akey specializes in what's known as rare variation, or changes in DNA that are found in perhaps one in a hundred people, or even fewer. For practical reasons, rare variants have been studied in earnest only for the last several years. Before then, it was simply too expensive. Genomics focused mostly on what are known as common variants.

However, as dramatically illustrated by a landmark series of papers published this year—by Alon Keinan and Andrew Clark, by Matt Nelson and John Novembre, and another by Akey's group, all appearing in *Science*, along with new results from the humanity-spanning 1,000 Genomes Project—common variants are just a small part of the big picture. They're vastly outnumbered by rare variants, and tend to have weaker effects.

The medical implications of this realization are profound. The previously unappreciated significance of rare variation could explain much of why scientists have struggled to identify more than a small fraction of the genetic components of common, complex disease, limiting the predictive value of genomics.

But these findings can also been seen from another angle. They teach us about human evolution, in particular the course it's taken since modern *Homo sapiens* migrated out of Africa, learned to farm, and became the planet's dominant life form.

"We've gone from several hundred million people to seven billion in a blink of evolutionary time," said Akey. "That's had a profound effect on structuring the variation present in our species."

Akey isn't the first scientist to use modern genetic data as a window into recent and ongoing human evolution, nor the first to root rare variation in humanity's post–Ice Age population boom. The new study's insights reside in its depth and detail.

The researchers sequenced in exhaustive detail protein-coding genes from 6,515 people, compiling a list of every DNA variation they found—1,146,401 in all, of which 73 percent were rare. To these they applied a type of statistical analysis, customized for human populations but better known from studies of animal evolution, that infers ancestral relationships from existing genetic patterns.

"There were other hints of what's going on, but nobody has studied such a massive number of coding regions from such a high number of individuals," said the geneticist Sarah Tishkoff of the University of Pennsylvania.

Akey's group found that rare variations tended to be relatively new, with some 73 percent of all genetic variation arising in just the last five thousand years. Of variations that seem likely to cause harm, a full 91 percent emerged in this time.

Why is this? Much of it is a function of population growth. Just ten thousand years ago, at the end of the last Ice Age, there were roughly 5 million humans on Earth. Now there are 7 billion. With each instance of reproduction, a few random variations emerge; multiply that across humanity's expanding numbers and enormous amounts of variation are generated.

Also playing a role are the dynamics of bottlenecks, or periods when populations are reduced to a small number. The out-of-Africa migration represents one such bottleneck, and others have occurred during times of geographic and cultural isolation. Scientists have shown that when populations are small, natural selection actually becomes weaker, and the effects of randomness grow more powerful.

Put these dynamics together, and the *Homo sapiens* narrative that emerges is one in which, for non-African populations, the out-of-Africa bottleneck created a period in which natural selection's effects diminished, followed by a global population boom and its attendant wave of new variation.

The result, calculated Akey, is that people of European descent have five times as many gene variants as they would if population growth had been slow and steady. People of African descent, whose ancestors didn't go through that original bottleneck, have somewhat less new variation, but it's still a large amount: three times more variation than would have accumulated under slow-growth conditions.

Natural selection never stopped acting, of course. New mutations with especially beneficial effects, such as lactose tolerance, still spread rapidly, while those with immediately harmful consequences likely vanished within a few generations of appearing. But most variation has small, subtle effects.

It's this type of variation that's proliferated so wildly. "Population growth is happening so fast that selection is having a hard time keeping up with the new, deleterious alleles," said Akey.

One consequence of this is the accumulation in humanity of gene variants with potentially harmful effects. Akey's group found that a full 86 percent of variants that look as though they might be deleterious are less than ten thousand years old, and many have existed only for the last millennium.

"Humans today carry a much larger load of deleterious variants than our species carried just prior to its massive expansion just a couple hundred generations ago," said the population geneticist Alon Keinan of Cornell University, whose own work helped link rare variation patterns to the population boom.

The inverse is also true. Present-day humanity also carries a much larger load of potentially positive variation, not to mention variation with no appreciable consequences at all. These variations, known to scientists as "cryptic," might be evolution's hidden fuel: mutations that on their own have no significance can combine to produce unexpected, powerful effects.

Indeed, the genetic seeds of exceptional traits, such as endurance or strength or innate intelligence, may now be circulating in humanity. "The genetic potential of our population is vastly different than what it was ten thousand years ago," Akey said.

How will humanity evolve in the next few thousand years? It's impossible to predict but fun to speculate, said Akey. A potentially interesting wrinkle to the human story is that, whereas bottlenecks reduce selection pressure, evolutionary models show that large populations increase selection's effects.

Given the incredible speed and scope of human population growth, this increased pressure hasn't yet caught up to the burst of new variation, but eventually it might. It could even be anticipated, at least from theoretical models, that natural selection on humans will become stronger than it's ever been.

"The size of a population determines how much selection is going to be acting moving forward," said the anthropologist Mark Shriver of Penn State University. "You have an increase in natural selection now."

An inevitably complicating factor is that natural selection isn't as natural as it used to be. Theoretical models don't account for culture and technology, two forces with profound influences. Widespread use of reproductive technologies like fetal genome sequencing might ease selection pressures, or even make them more intense.

As for future studies in genetic anthropology, Akey said scientists are approaching the limits of what can be known from genes alone. "We need to take advantage of what people have learned in anthropology and ecology and linguistics, and synthesize all this into a coherent narrative of human evolution," he said.

The geneticist Robert Moyzis of the University of California, Irvine, coauthor of a 2007 study on accelerating human evolution, noted that the new study looked only at protein-coding genes, which account for only a small portion of the entire human genome. Much of humanity's rare variation remains to be analyzed.

Moyzis' coauthors on that study, the geneticist Henry Harpending of the University of Utah and the anthropologist John Hawks of the University of Wisconsin, also warned against jumping to early conclusions based on the new study's dating. Some of what appears to be new variation might actually be old, said Hawks.

Even with these caveats, however, the study's essential message is unchanged. "Sometimes people ask the question, 'Is human evolution still occurring?'" said Sarah Tishkoff. "Yes, human evolution can still occur, and it is."

"PARALLEL UNIVERSE" OF LIFE DESCRIBED FAR BENEATH THE BOTTOM OF THE SEA

Deep beneath the ocean floor off the Pacific Northwest coast, scientists have described the existence of a potentially vast realm of life, one almost completely disconnected from the world above.

Persisting in microscopic cracks in the basalt rocks of Earth's oceanic crust is a complex microbial ecosystem fueled entirely by chemical reactions with rocks and seawater rather than sunlight or the organic byproducts of light-harvesting terrestrial and aquatic ecosystems.

Such modes of life, technically known as chemosynthetic, are not unprecedented, having also been found deep in mine shafts and around seafloor hydrothermal vents. Never before, though, have they been found on so vast a scale. In pure geographical area, these oceanic crust systems may contain the largest ecosystem on Earth.

"We know that Earth's oceanic crust accounts for 60 percent of Earth's surface, and on average is four miles thick," said the geomicrobiologist Mark Lever of Denmark's Aarhuis University, part of a research team that describes the new systems in the journal *Science*.

If what the researchers found resembles what's found elsewhere below Earth's oceans, continued Lever, "the largest ecosystem on Earth, by volume, is supported by chemosynthesis."

The paper represents the culmination of findings that have gathered over the last two decades, starting in the 1990s with the discovery of strange microscopic holes in the basalt rocks that form much of Earth's outer crust, floating above the planet's viscous upper mantle and below seafloor sediments.

The holes looked as though they were made by bacterial activity, but there wasn't supposed to be any life there. The crust isn't just hot, deep, dark and dense, but mostly devoid of the organic compounds, supplied by plants and plankton and other sunlight-fueled organisms, on which life relies elsewhere.

In coming years, researchers noted that oceanic crusts, which form when rock heated by Earth's core pours slowly through mid-ocean cracks between continental plates, differed greatly between the centers and edges. At the centers, near where they form, rocks are suffused with energy-rich compounds that support microbes. At the edges, where rocks arrive millions of years later, the chemicals are gone. It's like they've been eaten.

Other researchers found DNA traces of microbes in the oceanic crusts, further making the case for life, but just what the microbes were doing remained uncertain.

"All these pieces of evidence have been coming together for over 15 years. It was time to put it all together," said the microbial ecologist Andreas Teske of the University of North Carolina, a co-author of the new study. "We now have the best available evidence that there is in fact microbial life in the cracks and fissures of deep ocean basalt. The question is, how far does it extend?"

Teske and Lever's team collected samples of crust from the Juan de Fuca Plate, about 120 miles off the coast of Washington, drilling from boreholes made by other researchers some 1.5 miles below the ocean's surface and beneath another 1,000 feet of sediment.

At that depth, there exists rock and water and carbon dioxide, and few if any traces of organic matter produced from sunlight in the illuminated surface world. The researchers put their rocks in a laboratory apparatus designed to simulate that environment, then spent the next seven years observing what happened.

They measured chemical ebbs and flows, slowly learning the system's elemental cycles. Though the microbe populations didn't grow at the densities necessary to conveniently find them under a microscope, the researchers scoured their rocks for microbial DNA, identifying sequences that could be compared with known genes. Out of this emerged a picture of the oceanic crust community and how it lives.

Fundamental to the ecosystem is hydrogen, which in the absence of sunlight provides the energy on which all other biological processes rely. The hydrogen is released by reactions between iron- and sulfur-rich rocks and seawater, then used by microbes to fuel their conversion of carbon dioxide into organic matter.

That matter, along with metabolic byproducts like methane, would support other organisms, ultimately creating a web of life. That web is relatively simple compared with sunlight-based ecosystems, said Teske, and it's unlikely that multicellular life will be found there, as it's too hot and energy-poor compared with the places where higher life is found.

The work "confirms that there are subsurface environments that can support life without using oxygen," said Martin Fisk, a biogeochemist at Oregon State University who also studies oceanic crust microbiology at the Juan de Fuca Plate, but was not involved in the new research.

The biogeochemist Everett Shock of Arizona State University, also not involved in the research, isn't yet ready to rule out multicellular life. "My bet is on fungi," he said, "but there are other possibilities, including things that may be quite unfamiliar."

Continued Shock, "As for invertebrate and vertebrates, much depends on their size and the sizes of interconnected pore spaces in the rocks. I'm not ready to rule out such possibilities. Our ignorance about these systems is staggering, and accessing them is not at all easy."

Even if multicellular life isn't to be found in oceanic crusts, the presence of any life there is still extraordinary. Lever emphasized just how disconnected it is from the rest of Earth's life processes, a sort of "parallel universe" linked to ours only by seawater.

Despite that tenuous link, said Lever, it's likely that over geological time "those processes happening in the crust have a profound chemical influence on the composition of our oceans and atmosphere."

Another avenue of speculation involves the origins of life, which some scientists think might be traced to oceanic crusts. If simple interaction between seawater and rock provides the necessities, then Earth's early environments were quite conducive to life.

"The emphasis on common processes is appealing," Shock said. "It moves attention away from special circumstances, like spark discharges in implausible atmospheres, or conditions that may once have prevailed but no longer do."

Lever mused on the possibility that primeval chemical systems with a tendency to replicate themselves, not quite alive yet something more than inanimate, might have accreted around hydrogen- and sulfur-generating processes in the oceanic crust.

"What's proposed is that before there was life, there was this organic matter-producing chemical reaction going on," said Lever. Life might have originated around the iron and sulfur compounds fueling that reaction, evolving to produce biomass and harvest energy.

Such ideas are speculative, emphasized Lever, and Teske preferred to think about the implications for life elsewhere. "What I find interesting here is not so much the origin of life, but the persistence of life," he said.

"As long as there's space for microbes, and biochemistry, life persists," Teske continued. "Deep subsurfaces could be the best hiding place for life on other planets, where surface conditions are too harsh but the right chemical conditions are available below."

Back on Earth, a more immediate implication of the findings is the possibility that a large fraction of Earth's life exists in oceanic crusts, not in ocean water or on land.

"We need to stretch our brains to consider that there should be much to discover, and much that will be unfamiliar," said Shock.

AT THE EDGE OF INVASION, POSSIBLE NEW RULES FOR EVOLUTION

Just as Galapagos finches are icons of evolution by natural selection, Australia's cane toads may someday be icons of "spatial sorting"—a dynamic that seems to exist at the edges of invasion, altering the standard rules of evolution.

Cane toads have evolved in odd ways Down Under. Adaptations that drove their dramatic spread made individual toads less reproductively fit. Evolution through natural selection of hereditary mutations still exists, but no longer appears driven by reproductive imperatives alone. It's also shaped by speed.

"The possibility that some traits have evolved by 'mating betwixt the quickest' rather than 'survival of the fittest' warrants further attention," wrote biologists led by the University of Sydney's Richard Shine in the *Proceedings of the National Academy of Sciences.*

Introduced to northeast Australia seventy-five years ago in an ill-advised attempt at beetle control, cane toads spread like fire, their range expanding at rates that grew daily. When they first arrived at his study area, Shine noticed something strange: As expected, the toads displayed myriad adaptations—longer legs, greater endurance, a tendency to move faster and farther and straighter—that affected their ability to disperse, but dispersal's benefits were unclear.

The fastest-spreading cane toads also had the highest mortality rates. Longer, stronger toad legs led to spinal injuries. "Most obviously, why did the toads just sprint through our magnificent, food-rich flood plain in a frenetic rush to keep on going?" said Shine. After all, if the toads' evolution is driven solely by a drive to reproduce, they would have stopped to enjoy the spoils of invasion.

"Much of what they did seemed hard to reconcile with the idea of natural selection enhancing individual fitness," said Shine. "We started thinking about what other kinds of processes could have caused them to become such driven little robotic dispersal machines."

In the new study, Shine describes those processes, which fall under the rubric of "spatial sorting" and are most easily understood by analogy: Imagine a race between rowboats crewed by randomly distributed oarsmen. If the race is stopped intermittently, and oarsmen randomly redistributed between boats nearest one another, boats in the lead will accumulate ever-higher proportions of skilled rowers.

Those are the dynamics of spatial sorting. Boats are organisms, rowers are genes and the crew swap is reproduction. Each newly crewed boat is offspring. Generation by generation, organisms in the lead get faster and faster. Classical natural selection still operates—if a mutation causes an organism's offspring to go sterile, the lineage soon ends—but it's no longer the only driver of evolution.

Now space matters too. Physical proximity produced by dispersal continue to shape that dispersal. Whatever drives creatures to spread farther and faster clusters at the front. If an adaptation improves dispersal but hurts survival, it matters less than usual, because the pool of potential mates is determined by their ability to cover ground.

A key challenge in studying spatial sorting is disentangling its effects from those of natural selection. In many cases, better dispersal is a good, old-fashioned adaptation: It might help organisms find new sources of food or relieve overcrowding.

Such disentanglement is presently hard to do, wrote Shine. Cane toads are the best-studied candidate for spatial sorting, though gaps in data still exist.

But spatial sorting might help explain instances of a phenomenon called preadaptation, in which complex traits emerge through the combination of many smaller adaptations, each of which provides no survival advantages. It would seem unlikely for them to persist long enough to collect in one place—unless, that is, survival advantages are no longer so important. And in a world full of biological invasions, anything that helps explain their dynamics deserves further study.

"Spatial sorting may prove to be classical natural selection's shy younger sibling, not as important as Darwinian processes but nonetheless capable of shaping biological diversity by a process so-far largely neglected," wrote Shine's team.

A MUD-LOVING, IRON-LUNGED, JELLY-EATING ECOSYSTEM SAVIOR

Meet the bearded goby, a six-inch-long fish that lives in toxic mud, eats jellyfish, lasts for hours without oxygen, and has saved a coastal African ecosystem from a nightmare fate.

Over the last several decades, as other fish populations off the coast of the Namibia collapsed, jellyfish and bacteria populations exploded—a condition widely considered to be an ecological dead end, incapable of supporting rich webs of life.

But amidst this turmoil, the goby has thrived. It circulates nutrients that would otherwise be lost, feeds animals who lost their historic prey, and provides that rare thing: a happy, or at least not-so-bad, ending to an environmental disaster story.

The goby "has the ability to consume what were considered dead-end resources and convert them into bite-sized chunks for higher trophic levels," said Mark Gibbons, a University of the Western Cape biologist. "Gobies have become anything but a dead-end resource. The gobies are now sustaining the rest of the ecosystem."

Half a century ago, the bearded goby was just one of many species living in what's known as the Benguela Large Marine Ecosystem, about 7,000 square miles of continental shelf off the coast of southwest Africa.

The region supported a prosperous commercial fishing industry, but overfishing depleted the northern Benguela's keystone species, the sardine. By eating plankton and being eaten by larger fishes, sardines had provided a direct conduit between the bottom and top of the Benguela's food chain. Now that link was gone.

Adding to the upheaval, naturally occurring upwellings of deep, cold water in the Benguela deliver nutrient loads that feed enormous plankton blooms, which feed oxygen-gobbling, dead zone-creating bacteria and eventually fall to the ocean floor, forming a toxic sludge. Methane gathers in the mud, belching out in fish-killing gas eruptions. Without sardines to eat the extra plankton, the effects of this natural feature became more pronounced.

Such radical stresses produced what ecologists call a regime shift. The web of life didn't simply adjust a bit, but took a whole new form, one that didn't require a rich assortment of fishes to circulate energy and nutrients. In this lowest-common-denominator system, there were only a few opportunist fish species, bacteria and, at the top of the food chain, giant jellyfish.

Giant jellies have no natural predators, and aren't even eaten by humans. In the systems they dominate, nutrients and energy go from plankton to jelly, with little between. "The massive increase in jellyfish biomass after the collapse has been regarded as a trophic dead end," wrote Gibbons and colleagues in a study published in *Science*. The same has happened in China's Bohai Sea, the Sea of Japan, and the northwest Mediterranean. But unlike those ecosystems, the northern Benguela has the bearded goby.

In recent years, fishermen and researchers noticed more bearded gobies than before. Gobies were showing up in the bellies of seals, penguins, and the remaining large fish, such as horse mackerel and hake. But nobody quite knew what they were doing, so Gibbons, along with the University of Bergen biologists Anne Utne-Palm and Anne Salvanes, decided to find out.

They measured oxygen content and chemical composition throughout the northern Benguela's waters and across its floors. They used radar to track the movements of goby populations and conducted a series of aquarium experiments on individual fishes. What they found is a fish extraordinarily well-suited to its new environment.

During the day, gobies live on and in the Benguela's toxic sea sludge. They do fine without oxygen: After spending hours in aquariums filled with oxygen-free water, gobies are still alert. Given the choice between toxic mud and sand, they picked the sludge.

The gobies feed on the mud, scooping it up and waiting until evening, when they swim into the higher-oxygen water column, to digest it. While in the water column, they prefer to stay among giant jellyfish, whose stinging tentacles discourage predators from following. And the gobies have developed a taste for the jellies: The researchers' autopsies found that jellyfish can form up to 60 percent of a bearded goby's diet.

These adaptations are likely rooted in the gobies' evolution in the Benguela, where they dealt with toxic mud and low-oxygen waters, albeit in lower quantities than now, for millions of years. "This 'pre-conditioning' allowed them to capitalize on changes to the system," said Gibbons.

For many, the Benguela's current state is still far from ideal. Philippe Cury, a fisheries biologist at France's Institute of Research for Development, called it a "ghost ecosystem" for fisheries. "So tell your kid, 'Eat your gobies with your jellyfish!'" he said. But without the goby to feed other species—and, critically, to keep nutrients in circulation during particularly extreme years, when other fish can't survive—the situation would be far worse.

"There would be less hake, less seabirds, seals and cetaceans and all those other organisms that feed on gobies," said Gibbons. "That would be a desert."

Whether other jellyfish-dominated systems will prove to have their own versions of the bearded goby remains to be seen. But at least the northern Benguela has avoided total catastrophe.

"Fortunately for the Benguela, they had the goby," said Utne-Palm. "It's a lucky end to something that could have been more of a disaster."

REDEEMING THE LAMPREY

Several years ago, a young man bow-fishing on New Jersey's Raritan River spotted a long, thin creature in the murky water. He shot the animal through the neck, reeled it in, and posed for photographs. Eventually a friend posted one to Reddit. Within days it went viral, viewed 1.4 million times and setting the Internet ablaze with mystery-monster speculation. "Terrifying Sea Monster Found in New Jersey River," headlined Gawker. The *New York Daily News* likened the "large, unidentified creature" to "a scene from a sci-fi film." Some people insisted it was all a hoax.

The creature was, in fact, a common fish called a sea lamprey. The lamprey is admittedly unusual in appearance, with a boneless eel-like body and jawless tooth-ringed mouth, and unusually grotesque when blood-slicked and impaled on a stake. But they're also well-known residents of coastal waterways throughout eastern North America—which, to be fair, quite a few people pointed out. Yet that there was so much uncertainty, much less Sea Monster!-level horror, reflected the strange mix of ignorance and distortion through which sea lampreys are perceived.

Indeed, in certain circles, there's no fish more reviled. Ours is not a society kindly disposed to either predators or parasites, and sea lampreys are both: As adults they attach to other fish, using those toothy sucker-mouths to bore holes through which they siphon blood and bodily fluids. They're blamed for nearly destroying the commercial fisheries of the Great Lakes, where they've been subject to a century-long extermination program. Most fishermen and biologists see them as rapacious fish-killers, something to hate and fear.

And yet. Set aside the sea lamprey's reputation, at least for a moment. Heed the words of the biologist John Waldman's *Running Silver*, an account of the extraordinary richness of pre-industrial coastal river life: "Like so many outcasts," writes Waldman of the lamprey, "it is also profoundly misunderstood." Consider the possibility that even those creatures we most despise may be wellsprings of life, enriching the world in ways we've only started to understand. Or, to put it another way, "Anadromous sea lampreys (*Petromyzon marinus*) are ecosystem engineers in a spawning tributary."

That's the title of a paper published last year in the journal *Freshwater Biology*. Among the coauthors is Stephen Coghlan, a University of Maine fisheries ecologist who has spent much of the last eight years studying one Sedgeunkedunk Stream, a tributary of Maine's great Penobscot River. There the ongoing removal of several large dams has earned national attention, sparking hopes that the Penobscot's waters might again, per Waldman, run silver with migratory fish.

The same applies on a much smaller scale to Sedgeunkedunk, where an old dam's dismantling in summer 2009 made it a microcosm of the river system's potential recovery. Researchers view Sedgeunkedunk as a living laboratory, a before-and-after experiment to see what happens when long-inaccessible waters are reopened. In short: Fish come back. Alewives, a foundation-of-the-food-chain fish that once swam from the Penobscot to the Atlantic Ocean and back in the billions. Atlantic salmon, iconic and beloved and now almost absent from the river, once so common they were served in poorhouses. And sea lampreys.

Before coming to Maine, Coghlan worked in western New York, in the Great Lakes watershed, where the conventional American wisdom on sea lampreys was conceived after the late nineteenth- and early twentieth-century collapse of world-renowned trout populations. The lampreys were invasive, having infiltrated the region through shipping canals; Great Lakes fish could not adapt to their ravages. Historical records describe the ease with which nets could be lowered to lake bottoms, then pulled back up with a cargo of sea lamprey-holed carcasses. "They were the scourge," says Coghlan. Great Lakes fish are now sustained by a constant, watershed-scale campaign of poisoning the streams where the invaders spawn.

In the middle of the last decade, though, a series of studies led by John Waldman challenged that simple narrative. By comparing the genomes of the Great Lakes and Atlantic Seaboard lamprey, Waldman and colleagues were able to reconstruct their population histories—and the resulting family trees suggested that sea lampreys were in fact native to Lake Ontario, the easternmost of the Great Lakes, as well as Lake Champlain and the nearby Finger Lakes, where they have also been, erroneously, considered invasive.

Yes, at some point sea lampreys swam from Lake Ontario through the Welland Canal and into the other Great Lakes, and there caused havoc. That's indisputable. Yet in some places, they were not invaders, but instead had been present for thousands of years. Which raised the question: If sea lamprey are so terrible, and lived there all along, why wasn't Lake Ontario a lamprey-dominated wreck to begin with? Why did its once-flourishing fisheries collapse only recently?

An alternative explanation, says Coghlan, is that Lake Ontario's lamprey explosion was simply the last in a chain of ecological insults, and made possible by those that preceded it. European settlers had harvested the lake's fish with their usual foresight and regard for nature's limits—which is to say, none at all. They built dams across streams and creeks, blocking trout and salmon from their spawning grounds. The dams powered sawmills that processed timber cleared from surrounding countryside. Subsequent soil erosion clouded lake and stream waters. Mills and forest-replacing farms added yet more pollution, creating conditions unfavorable to other fish but readily tolerated by lamprey.

Perhaps most destructively, alewives got into the system, likely introduced as a food source for commercially favored fish. They soon outcompeted native small fish, so that trout and salmon had little else to eat—but alewives, it turns out, contain high concentrations of thiamine, which causes nervous system damage and birth defects in predators who consume too many. In coastal systems, where predators co-evolved with alewives and there's greater dietary variety, it's not a big problem. In the Great Lakes, it was a toxic recipe.

"Perhaps when trout and salmon populations were robust, when they had access to clean water and spawning habitat, they could withstand some amount of lamprey predation," says Coghlan. "But now there was this shift. Maybe with all these man-made changes, lamprey just flourished." Their exploding numbers, then, were only a consequence of ecological upheaval, not its cause. Far from one-dimensional villains, they're a sort of victim.

Or so the explanation goes. Coghlan is quick to note that it's just a hypothesis, albeit a very plausible one supported by ever-more evidence. Regardless, sea lampreys remain widely persecuted today, the original Great Lakes narrative being internalized far and wide. Yet at least a few biologists have come to question whether, particularly in coastal regions, they're indeed so destructive.

At Sedgeunkedunk, Coghlan wasn't tasked with killing lampreys. He was just there to study them, beginning with their migratory return each spring in May or June, swimming upstream via the Penobscot from an adulthood spent in the Atlantic Ocean. They might be fifteen or twenty years old, and these are their last weeks. Already their eyes and digestive systems are starting to break down. They've enough strength left for one final burst of mating and, crucially, nest-building.

The nests consist of a mound of rocks—some the size of softballs, pulled from surrounding streambeds and thrashed into place—and, on the downstream side, a long pit into which eggs are laid and fertilized. As described and empiricized in that *Freshwater Biology* article, all this digging and dragging dislodges small insects and other aquatic invertebrates, which become food for larger insects, and also for salmon that line up behind the spawning lamprey, eating dim-sum style.

Above the nest, the stream's hydrology also changes. Water flows slowly over the pit but fast over the mound, clearing it of silt. Pockets and hollows between pebbles in the mound become prime habitat for midges, caddisflies, mayflies—foundational creatures. Lamprey nest-building doesn't seem to increase the total population of these insects; rather, it *concentrates* them, creating comparatively energy-rich regions where insect-eating fish can feed at high efficiency, packing on as much nutrition as possible before their own migratory journey out to sea.

If that's a fine point, it's more obvious what happens when the lampreys finish spawning. They die. Their bodies fall to the streambed, where they decompose and are consumed—rich nutrient bursts, delivered from the deep sea. In one Massachusetts stream, a few hundred sea lamprey corpses provided a full 20 percent of the system's total phosphorus, and more than one ton of minerals and nutrients. Just as important is timing: Nests are built and bodies fall just as trees grow out overhead, their canopies blocking sunlight from reaching the stream—which in turn stifles the growth of photosynthetic algae, a decrease that ripples through the food web.

"From spring to summer the energy balance of the stream goes from higher energy demand to lower energy availability," explains Coghlan, and the lamprey help tilt it back. One can imagine lines of energy radiating outward from them, into insects and fish, and then all the creatures who eat them, out into the river to the deep Atlantic Ocean and back again, a perpetual feedback loop of life.

Might sea lampreys play a similar forest-vivifying role as migrating salmon once did on the Pacific coast? They're likely not quite that powerful, says Coghlan, but they're still very important. Keystone species, even, with a disproportionately large influence on their communities of life, like wolves or elephants or beavers.

All this is coming to light at a moment when dams are finally being removed and rivers long constricted restored to life. In this effort, people have focused on the most charismatic fish, the salmon and shad and alewives. Yet perhaps the sea lamprey's unappreciated efforts are necessary for those other fish to thrive. Rather than destroying life, they nourish it. Which raises one more question: How many other creatures, loathed like sea lampreys or simply unnoticed, in fact make our world a far richer place?

DECODING NATURE'S SOUNDTRACK

One of the most immediately striking features about Bernie Krause is his glasses. They're big—not soda-bottle thick, but unusually large, and draw attention to his eyes. Which is ironic, as Krause's life has been devoted to what he hears, but also appropriate, since it's the weakness of his eyes that compelled Krause to engage with sound: first with music, and later the music of nature. Nearsighted and astigmatic, Krause has spent most of the last half-century recording biological symphonies to which most of us are deaf.

Even more than Krause sees, he listens. It's an unusual trait. Ours is a society that privileges vision—illumination is insight, leaders are visionaries, do you see what I'm saying—and the habit extends to our interactions with nature. Take a walk in the woods and you'll almost certainly be soothed by birdsong and trilling streams and wind whispering in treetops. But when you get home, you'll likely be asked: What did you see?

That will likely be what you remember, too. The sounds will have been background noise. And in that habit, you won't be alone. In some ways, scientists have tended not to hear nature. Not that they've been deaf; there are plenty of studies on, say, sound production in croaker fish or the neurological basis of finch songs. But to Krause and the scientists he's inspired in the emerging field of soundscape ecology, traditional bioacoustics has too narrow a focus.

Their microphones are stethoscopes pressed to Mother Nature's chest. Krause records the feedback between the natural world and us.

"The voice of the natural world informs us about our place in the living world and how we're affecting it," he says. "It tells us everything we need to know about how we're doing in relationship to it."

At this particular moment in Earth's history—the morning of what some scientists call the Anthropocene, an age in which human influence on natural processes is ubiquitous and immense—we have many tools to measure our ecological impacts: by eye, generally, focusing on particular species or guilds of interest, counting them in the field, peering by satellite at changes in land use, and translating our observations into the language of habitat type and biodiversity.

To Krause, these are measurements best made by listening to natural soundscapes. In a career of listening and recording, he's amassed a veritable Library of Alexandria of nature's sounds, and he emphasizes that they're not merely recordings of individual creatures. The traditional approach of bioacoustics, focusing on single animals and species, is anathema. It's "decontextualizing and fragmenting," he says, like trying to extract a single violin from Beethoven's Fifth Symphony. "Take an instrument out of the performance, and try to understand the whole performance, and you don't get very much," he says.

Inevitably Krause has captured the players—bearded seals with voices that echo geomagnetic storms, baboons booming in granite amphitheaters, a fox kit playing with a microphone—but they're incidental to recording whole habitats and communities.

In his home studio, perched on an oak-covered hillside in Glen Ellen, California, Krause plays me some of his favorites: a Florida swamp, old-growth forest in Zimbabwe, intertidal mangroves in Costa Rica, and a Sierra Nevada mountain meadow. As the sounds pour from speakers mounted above his computer, spectrograms scroll across the screen, depicting visually the timing and frequency of every individual sound. They look like musical scores.

In each spectrogram, Krause points something out: No matter how sonically dense they become, sounds don't tend to overlap. Each animal occupies a unique frequency bandwidth, fitting into available auditory space like pieces in an exquisitely precise puzzle. It's a simple but striking phenomenon, and Krause was the first to notice it. He named it biophony, the sound of living organisms, and to him it wasn't merely aesthetic. It signified a coevolution of species across deep biological time and in a particular place. As life becomes richer, the symphony's players find a sonic niche to play without interference.

"The biophony is the pure expression of life, of the given organisms in a habitat," he says. "When you're in a healthy habitat, all the species are able to find bandwidth where their voices fit." He puts an ancient Borneo rain forest onto the speakers. At the top of the spectrogram are bats, their echolocation a bare hint of a sound to human ears; below them are cicadas, a plenitude of insects, one chestnut-winged babbler and nightjars and the booms of gibbons, each in its own place.

Krause magnifies the view, zooming in so that each animal's call can be discerned as individual orange dots. From this view, the spectrogram looks like constellations seen through a telescope. "You've got a whole universe in there," he says. "Look at the discrimination here. It's so beautiful. It tells you how old this habitat is."

It took a while for Krause's observation, initially made informally in the early 1980s, to reach sympathetic scientific ears. Krause was not, after all, a member of the scientific community. On his studio walls are photos from his early, pre-nature music career—Krause singing tenor with The Weavers' during a 1963 Carnegie Hall performance, the Moog synthesizer he played in early electronic music and for special effects in *Rosemary's Baby* and 1978's *Invasion of the Body Snatchers*.

But the idea of biophony struck a chord with Stuart Gage, an ecologist at Michigan State University interested in using sounds to take the pulse of ecosystems and monitor their ecological health. He and Krause recorded Sequoia National Park, using an early gauge of acoustic complexity, measuring patterns of species diversity across different habitats and seasons. It was a proof-of-principle test, and in recent years, with the advent of inexpensive recorders and big hard drives that make it possible to record large areas over long periods of time, the field—now dubbed "soundscape ecology"—has flourished.

Researchers are now recording dozens of landscapes across the world and computationally translating thousands of hours of recordings into numerical indexes of ecological conditions, and ultimately the human impacts on them. These are not ready for practical application. The biophony, formally known as the "niche hypothesis," is still a hypothesis. It's not yet possible to hear whether an ecosystem is healthy. But that may happen soon. "I think we're going to see a rapid shift into applied research," says Bryan Pijanowski, a Purdue University soundscape ecologist, whose projects include sonic assessments of a wildfire-scorched Sonoran desert.

Some of the current indexes—Gage is reviewing six, and there are many more—are relatively straightforward counts of animal calls or the total species in an area. Others, however, resonate more closely with Krause's notion of the "biophony." Jérôme Sueur at the National Museum of Natural History in France is trying to quantify species diversity. One of the indices his group developed, the Acoustic Entropy Index, is predicated on the notion that communities of life in undegraded habitats become not just diverse but also structured, like Krause's old-growth Borneo forest biophony.

The ecologist Almo Farina at Italy's University of Urbino and the Italian composer David Monacchi are applying similar ideas to other forests in Borneo, with Farina's student Nadia Pieretti leading studies of central Italian forests. When the patterns of birdsong in those forests are analyzed and turned into mathematical measures of complexity, says Pieretti, the symphonies of communities subject to road-building and intrusion indeed seem to be less structured. Birds call louder and repeat themselves, perhaps to be heard above vehicular din; there's more noise, but not more information.

This particular type of biophonic impact, in which human noise—the anthrophony—drowns out nature players, is one to which Krause is especially sensitive. It's been an active field of traditional bioacoustic study, with researchers finding many cases in which sound pollution seems to interfere with animal communication. That's not always the case, and often it's been difficult to disentangle the effects of human presence—the roads themselves, say, not motor noise—from those of our sounds. Soundscape ecology approaches noise pollution in more nuanced ways, focusing not only on noise's effects on particular species but across habitats and communities.

Although sound indexes of ecological health may be years away, pending rigorous testing, calibration, and codification, Krause says he doesn't need to wait for the results. He estimates that nearly half of the habitats he's recorded are now compromised or rendered silent, primarily by human development and insatiable appetites that relegate most nonhuman interests to irrelevance. Krause requires no scientifically validated tools to hear that feedback. "If you know how to listen to it, then it's really clear what's happening," he says. "As the natural world becomes more silent over time, the question is: Is that what we want?"

Krause is a father of soundscape ecology, but the methods of its maturity are not his own. He created the niche hypothesis and has made important contributions to the emerging science, but he's not going to blanket a forest with recorders and let his computer do the listening. There's something profoundly important to Krause about the human act of listening, about being there.

On a cool March morning, Krause drives me to one of his favorite spots, one he's recorded since the early 1990s: Sugarloaf Ridge State Park, 4,500 mostly undisturbed acres in the mountains between the Napa and Sonoma valleys. We pull into the parking lot and head up a trail before dawn. Above a ridge to the south, Antares blazes red; Venus incandesces like a flare.

The park is home to an observatory where local stargazers gather for celestial events. On the path, signposts mark the relative distance of the planets from the visitor center's sun. A little ways beyond Saturn and a tall willow where wild turkeys sometimes roost, beside a stand of coyote brush and bay laurel, Krause stops to set up his equipment. Working by the light of a headlamp, it takes only a minute. He probably could do it with his eyes closed.

Behind the brush a stand of alders and willows marks the edge of Sonoma Creek, burbling from the rains that ended the region's seven-month-long dry spell. It's good songbird habitat, and a few minutes later, after we've walked away from the recorder and up the trail, as the sky grows a shade brighter, they start to sing.

It's the beginning of the dawn chorus: that universal performance, whether in rain forest or tundra or this particular semi-arid riparian pocket, when the community of life greets a new day. Most of Krause's work, and much of soundscape ecology recordings, involves recordings taken at this information-rich time.

Just why the chorus happens, though, remains a mystery. Many explanations have been proffered, from territorial display to the unique sonic qualities of morning air to the possibility that, in such dim light, there's nothing else for birds to do. Farina and Pieretti think the creatures in a biophony aren't just talking to others of their own kind, but listening to each other. Nature recorder Martyn Stewart calls it "a newspaper being read in dozens of different languages."

First to read the news is a scrub jay. In a few minutes he's joined by white-crowned and song sparrows, titmice and juncos and towhees and mourning doves. A red-shouldered hawk is heard, and a pileated woodpecker, and a great-horned owl. Just before we return to the recorder, the turkeys make their appearance. They've got a lot of news.

Back in the studio, Krause seems a little disappointed, almost like someone who's heard a band so many times that only exceptional performances stand out. Compared with other choruses he's heard at that spot, ours was relatively quiet—reflecting, he suspects, the lingering consequences of drought, or changes in local climate. The last few years, says Krause, seem to have become quieter.

Despite Krause's misgivings, it's still an extraordinary concert, at least to my ears. Above the creek's burble, the birds have arrayed themselves neatly in sonic space. The chorus is still rich and orderly. A few jets can be heard in the distance, but without causing any apparent disturbance. Sugarloaf is far from pristine—it was ranched a century ago—but it's now protected and generally undisturbed. Despite the drought and surrounding urban sprawl, there's still habitat, a chance for life to flourish.

"Although it is not nearly as robust as in past years, either in density or diversity," says Krause, "it is, nonetheless, a sonorous and hopeful expression of seasonal life." As much as Krause laments that many soundscapes he's recorded will never be recovered, in more hopeful moments he talks of those that might still be protected, even regrow. Where we allow it, life will thrive. If only we would listen.

Author's note: Since I visited Krause, the drought has continued unabated. He has continued to record at Sonoma Creek, and in March 2016 published with Almo Farina a study in the journal *Biological Conservation* describing the biophony's long-term progression. "The results indicate a sharp decrease of acoustic complexity of biophonies associated with a decrease of geophonies from a stream since 2011," they wrote. As the stream has dried, life has dwindled.

II

INNER LIVES

One of the first times I really thought about what animals experience occurred while reporting on researchers trying to understand how octopi coordinate the movement of their eight appendages.

It's quite the feat. After all, scientists struggle to program robots to control a mere two arms. And to learn more, the researchers inserted fine wires into their octopi's brains, administered an electric current, and then cross-referenced the resulting movements with the location of stimulated brain regions. From that reverse-engineered neurological map they learned many interesting things: that different brain regions controlled different body parts at different times, and that the neurological programs are stored not only in the octopi's brains but also in their arms.

The findings, said the researchers, could be used to design better robots. Yet I kept thinking of the octopi themselves. What did they make of their lives?

I'd considered animal experience previously, to be sure—loved my pets, had my heart melt while watching cute videos of prairie dogs, and so on. But I'd internalized the rules by which modern society views animals: some we acknowledge whereas others are invisible, and though in an everyday way we admit their intelligences and experiences, in the scientific context they're depersonalized.

The electrically stimulated octopi made me confront that, and in coming years I devoted much more of my attention to the science of animal minds. This coincided a flourishing of such research; hardly a week goes by now without some headline-grabbing demonstration of animal thoughtfulness. It's left me with the conviction that to walk outside is to walk into a kaleidoscope of inner worlds.

The poet and environmentalist Gary Snyder described ecological relationships as "nets of beads, webs of cells." Through his eyes, landscapes pulsate; each organism is suspended in a vast, interconnected field of energy and life, glowing like so many illuminated jewels. To this I would add: each of those jewels is thinking too.

BEING A SANDPIPER

I met my first semipalmated sandpiper in a crook of Jamaica Bay, an overlooked shore strewn with broken bottles and religious offerings at the edge of New York City. I didn't know what it was called, this small dun-and-white bird running the flats like a wind-up toy, stopping to peck mud and racing to join another bird like itself, and then more. Soon a flock formed, several hundred fast-trotting feeders that at some secret signal took flight, wheeling with the flashing synchronization that researchers observing starlings have mathematically likened to avalanche formation and liquids turning to gas.

Entranced, I spent the afternoon watching them. The birds were too wary to approach, but if I stayed in one spot they would eventually come to me. They followed the tideline, retreating when waves arrived, and rushing forward as they receded, a strangely affecting parade. When they came very close, their soft, peeping vocalizations enveloped me. That night I looked at photographs I'd taken, marveling as the birds' beauty emerged from stillness and enlargement, each tiny feather on their backs a masterpiece of browns. I looked up their scientific classification, *Calidris pusilla*, conversationally known as the semipalmated sandpiper—a name derived from a combination of their piping signal calls and the partially webbed feet that keep them from sinking in the tidal sand flats of their habitat, where they eat molluscs, insect larvae, and diatom algae growing in shallow, sun-heated seawater.

I learned that semipalmated sandpipers are the most common shorebird in North America, with an estimated population around 1.9 million. My copy of *Lives of North American Birds* described them as "small and plain in appearance," which seemed unappreciative, especially in light of their migratory habits. Small enough to fit in my hand, they breed in the Arctic and winter on South America's northern coasts, flying several thousand miles each spring and fall, stopping just once or twice. The flock I'd watched was a thread in a string of globe-encircling energy and life, fragile yet ancient, linking my afternoon to Suriname and the tundra. At that fact, I felt the sense of wonder and connection that all migratory birds inspire. Yet not once did I wonder what they thought and felt along the way. How did they experience their own lives, not just as members of a species but as individuals?

It was a question outside my habits of thought, and occurred to me only months later, when I interviewed the American artist James Prosek. His compendium *Trout: An Illustrated History* had earned comparisons to the great American ornithologist and painter John James Audubon. Prosek's paintings are indeed beautiful, and his book, published while he was still an undergraduate, was shaped by a tradition of field guides and natural histories.

Prosek had not personally encountered many of the trout and salmon species that he painted. Instead, he accepted on faith their place within established taxonomic classifications. But those classifications would soon be rearranged by the application of molecular genetics to the taxonomy of the salmonids, a rearrangement that encouraged Prosek's deepening appreciation for how varied fish of the same species or subspecies, or even the same watershed, could be. The field guide notion of a species "type" felt inadequate, even misleading. Prosek's contemplations culminated in the glorious paintings of his latest book, *Ocean Fishes*: he made a simple but profound decision to paint the specific, individual fishes he encountered. The Linnaean system of classification—a hierarchical naming structure introduced by the Swedish botanist Carl Linnaeus in 1735—might describe the world and its generalities, implied Prosek, but it could not capture the richness of an individual life.

Several months after meeting Prosek, I was walking in Jamaica Bay on a bitterly cold and cloudless day when I saw semipalmated sandpipers again, running ahead of a pounding surf that caught the afternoon sun and sprayed their retreats with prisms. As Elizabeth Bishop observed in her poem "Sandpiper": "The roaring alongside he takes for granted,/and that every so often the world is bound to shake." I wondered what it would be like to be one of them, to run with the flock and feed in the surf, to experience life at their scale and society. Simply put, did they enjoy it? Were they cold? Did they remember their journeys, feel a connection to individuals with whom they'd flown, a concern for compatriots and mates?

Asking those questions made me appreciate just how deeply I'd internalized the taxonomic system against which Prosek strained, as well as the habit of explaining animal behavior in mechanical terms. I'd regarded the sandpipers as embodiments of their species and life history, but not as individuals, much less as selves. This oversight was not coincidental. The very history of taxonomy and attendant studies of animal behavior is intertwined with a denial of individual animal consciousness.

Scientific taxonomy began not in the eighteenth century with Carl Linnaeus, but some two thousand years earlier in ancient Greece with philosophers who venerated rationality and the power of language. Before them, and especially during humankind's long prehistory, animal deaths at our hands might have been necessary or justifiable, but they were also seen as unfortunate. We offered thanks and apologies, as evidenced in paintings, artifacts, and ritual.

The most rationalistic of Greek thinkers washed their hands of such sentiments. Aristotle introduced the notion of "binomial nomenclature," grouping animals by whether or not they had blood, and whether they lived on land or in water, in a hierarchy with humans at the top. In his view, animals were incapable of any sensations but pain and hunger. Brutal as this sounds, Aristotle was practically an ancient Peter Singer compared with Stoics such as Zeno of Citium, who insisted that animals felt nothing at all. This view influenced early Christian thought and, eventually, René Descartes, according to whom animals were all body and no mind, no different from the lifelike mechanical toys popular in seventeenth-century France.

Descartes' influence is manifest in the infamous words of the French rationalist Nicolas Malebranche, who said in *The Search after Truth and Elucidations* that animals "eat without pleasure, cry without pain, grow without knowing it; they desire nothing, fear nothing, know nothing." Not everyone agreed. Notable critics included Thomas Hobbes, Spinoza, and Voltaire, but their objections held little sway in an era of triumph for mathematics and the physical sciences. It was an intellectual moment most unfavorable to what could be felt but not quantified. Thus beliefs about animals that would be considered psychopathic if acted out by a twenty-first-century child became tenets of Western scientific thought and, in this milieu, taxonomy as we know it took form.

The science of taxonomy was driven by wonder and new discoveries in faraway lands, but this was not the whole of it. As Michel Foucault notes in *The Order of Things*, people had always been interested in plants and animals. What taxonomy satisfied was not simply curiosity but a desire for an overarching order to the world. Linnaean classification was triumphant among dozens of competing, lesser taxonomical schemes, but they all served a common project of bringing nature's wild diversity to Enlightenment heel, of putting the messy living world in tabular form.

Linnaeus did have an extraordinary eye for detail, and combined his supreme ambition with a simple and powerful system for classification. It worked by comparing a few clearly visible, easily measurable anatomical traits: his natural history was based purely on surfaces. A century later, the French naturalist Georges Cuvier revolutionized taxonomy by introducing comparisons of internal anatomy, but, as far as the inner lives of animals went, this too was a superficial revolution. It was a science of gross anatomy, not of minds, reflecting Descartes's mechanistic image of animals as assemblages of pieces. By the time of Cuvier, science had an entrenched species-first filter through which nature would be scientifically and culturally apprehended.

Taxonomic science was, to be sure, far from arbitrary. It was, and is, a wonderful means of describing the variations that do exist in the natural world. Taxonomy—or modern-day systematics—provides a language with which it is possible to understand the sandpipers in that crook of Jamaica Bay as being part of a related group including oystercatchers and common terns. With that language, it's also possible to note that semipalmated sandpipers can live for more than a decade, take mates in monogamous relationships that may persist for years, eat a lot of horseshoe crab larvae while migrating, and have declined in population by roughly one-third since the 1980s.

Most important, taxonomy was a scaffold on which evolutionary theory could be built. Although Linnaeus had believed the variation among animals was an immutable arrangement and divinely apportioned, evolutionary thinkers realized that these were family resemblances, to be elucidated more than a century later by Charles Darwin. And the great beauty of evolution, its essential profundity, is in placing humans among animals, not only in body but also in mind.

Just as humans shared physical traits with other creatures, Darwin argued, so we also shared mental traits. The ability to think and feel was just another adaptation to life's uncertainties and hazards, and, given our evolutionary relatedness to all other living things, it made no sense for them to be unique to us. "Even insects express anger, terror, jealousy, and love," he wrote in *The Expression of the Emotions in Man and Animals*. His protégé George Romanes, who was an avid collector of anecdotes about intelligent cats and dogs, thought that animal behavior should be interpreted in light of our own capacities. "Whenever we see a living organism apparently exerting an intentional choice," wrote Romanes in *Animal Intelligence*, "we might infer that it is a conscious choice."

By emphasizing the kinship between animals and human beings, Darwinian taxonomy could have opened the door to thinking about the consciousness of individual animals. But, instead, the opposite happened. Even as evolution's mechanics were accepted and expanded, the views of Darwin and Romanes on individual animal consciousness were rejected, consigned to cautionary tales of how even the most brilliant scientists can get things wrong. By the 1940s, when the great systematist Ernst Mayr settled on a fuzzy but useful standard definition of a species—as a population with a common reproductive lineage that could interbreed—the possibility of animal consciousness and individuality, so evident to anyone with a pet dog or cat, was largely eliminated from mainstream science. We could accept our animal bodies, and classify ourselves on that basis, yet had to avoid the implication that animals might have human-like minds.

A new age of machines and industry spawned the behaviorism of the psychologist B. F. Skinner, who, echoing Aristotle and Descartes, proposed that animals were nothing but conduits of stimulus and response (as were humans). Seeming evidence of higher thought was an illusion produced by some simpler mechanism. It's true that behaviorism helped to establish protocols by which animal cognition could eventually be studied in rigorous, scientifically acceptable fashion. But the price was steep: decades would pass before scientists began to allow that some animals might be more than biological automata.

In the 1960s, Jane Goodall was mocked by her primatologist peers for speaking of chimpanzee emotions, such as a mother grieving for her dead infant. Even her use of gender-specific terms for individual chimpanzees was seen as anthropomorphic and unscientific. As the journalist Virginia Morell recounts in *Animal Wise*, Goodall's editor at the prestigious journal *Nature* tried to replace the pronouns "he" and "she" with "it" in her first manuscript. When the zoologist Donald Griffin wrote in *The Question of Animal Awareness* that biologists should investigate "the possibility that mental experiences occur in animals and have important impacts on their behaviours," it was still a radical suggestion.

These days, Goodall is a hero, Griffin a prophet, and studies of animal intelligence ubiquitous: not just in chimpanzees, dolphins, and parrots, but in octopuses, archerfish, prairie dogs, and honeybees as well—a veritable Noah's Ark of braininess. Caveats remain, of course. Intelligence is relatively easy to study, but it isn't quite the same thing as consciousness or emotional life. It's been less controversial to ask whether rats remember where they stored food than whether one rat cares for another.

Yet even rats, it turns out, feel some empathy for one another. A team at the University of Chicago found that rats became agitated when seeing surgery performed on other rats, and a follow-up study observed that, when presented with a trapped labmate and a piece of chocolate, rats free their caged brethren before eating. Those who study animal behavior are still careful when talking about subjective experience—sure, Eurasian jays can guess what their mates want to eat, but who knows if they like each other?—but they're being professionally cautious rather than dismissive. The average person can safely speculate away: animal consciousness is a reasonable default assumption, at least for vertebrates, and not just in some dim sense of the word, but possessing forms of self-awareness, empathy, emotion, memory, and an internal representation of reality.

Many of the characteristics thought to be important for higher consciousness (such as brain size) and a sense of individuality (in humans and—maybe, just maybe—a few other great apes and cetaceans) aren't so unique anymore, or are no longer considered very important. Features such as working memory and episodic memory—keeping multiple pieces of information in mind and remembering what has happened, the cognitive fundaments of conscious experience—appear to be widespread. And the environmental challenges that might prompt the evolution of consciousness are widespread too. Among these is sociality: if you're going to live with others, it's very useful to be conscious of them. And the distinction between cognition and emotion is increasingly seen as a false one: certainly in humans they are more or less inseparable systems.

In July of 2012, a group of high-profile neuroscientists signed the Cambridge Declaration on Consciousness with the announcement: "The weight of evidence indicates that humans are not unique in possessing the neurological substrates that generate consciousness. Non-human animals, including all mammals and birds, and many other creatures, including octopuses, also possess these neurological substrates." Those other creatures likely include a great many reptiles, amphibians, and fish. They tend to be underappreciated because they're even harder to study than mammals and birds and octopuses, and perhaps strike us as a bit inscrutable. Consciousness is necessary to be an individual, to have unique thoughts and feelings rooted in one's own experience of life—and the animal kingdom teems with it.

Many scientists still don't know this, or don't accept it. Take the whale biologist Shane Gero, who is part of a research team that has conducted long-term sperm-whale studies off the island of Dominica in the Caribbean: these studies describe the dynamics of whale families in which children are, in a very real sense, the center of their lives. Yet Gero told me of being chastised by colleagues for referring to animals by name rather than number. Pressure still exists to think not of individuals but of general species traits that happen to be manifested in a particular animal.

Gero has helped to decode the vocalizations that sperm whales might use as names, something that's also been observed in dolphins, but this remains controversial. That's why a visitor to the "Whales: Giants of the Deep" exhibition at the American Museum of Natural History in New York could learn a lot about their skeletons, heart capacity, and navigational abilities, but barely anything about their intelligence and social lives—arguably the most dynamic area of contemporary cetacean research.

Not surprisingly, the science of animal personality is still young. Recognition of animal consciousness might be just a first step. Individual differences based on temperament and experience, again so obvious to pet owners, is a new idea in science. For sperm whales, says Gero, such differences were once dismissed as statistical noise or evidence of behavioral maladaptation. The blind spot is hardly restricted to whales. The article "Energy Metabolism and Animal Personality," published in the journal *Oikos* in 2008, pointed out that "personality will introduce variability in resting metabolic rate measures because individuals consistently differ in their stress response, exploration or activity levels." Animals that have "frozen" with fear during capture might thus be misclassified as having high resting metabolic rates when in fact a motionless rabbit with his heart racing might simply be scared.

This seems like common sense—and in some respects the general public outpaces much of the scientific community, at least when it comes to the familiar animals we live with and know well. All those cute cat videos, reliably mocked as a symptom of our unintellectual Internet habits, bespeak our era's willingness to acknowledge the inner lives of companion animals. Not that they're tiny humans in kitten suits, of course—indeed, part of the fun in knowing a cat (not to mention watching those videos) is the obvious disparity between their view of the world and our own. But neither are they entirely incomprehensible, per Ludwig Wittgenstein's oft-quoted statement: "If a lion could speak, we would not understand him." Wittgenstein probably never saw a pair of lion cubs at play.

What might it mean to treat all vertebrates as having some form of consciousness and individuality? Animal welfare advocates campaign for the better treatment of companion and farm animals, which is a noble cause. But I am more interested in wild animals, our neighbors in nature. To the painter James Prosek, seeing wild animals as individuals offers a new and sorely needed conservation ethos. Biodiversity and ecosystem services make for well-meaning but often uninspiring rhetoric; they value nature generally, but provide little reason to care for actual creatures in a nearby forest or your backyard. Acknowledged as individuals, those sparrows, salamanders, and squirrels are not interchangeable parts of a species machine. They are beings with their own inner lives and experiences.

Does this mean we should never eat a salmon or cut down a tree to build a house? Not necessarily. We might simply acknowledge the consequences of our actions and offer apologies and thanks to those creatures we affect. It's the sort of ethical equation people need to solve for themselves.

For myself, I'd be happy to see a revival of naturalist language, the sort of charming, unapologetically anthropomorphic descriptions one finds in old field guides, written before the ascendance of the twentieth-century's airless, specialist vernacular. It's a voice heard in *The Birds of Essex County, Massachusetts*, in which Charles Wendell Townsend described a "low, rolling gossipy note" voiced by semipalmated sandpipers approaching other birds. He waxed eloquent about their courtship, the male "pouring forth a succession of musical notes, continuous wavering trill, and ending with a few very sweet notes that recall those of a goldfinch . . . one may be lucky enough, if near at hand, to hear a low musical cluck from the excited bird. This is, I suppose, the full love flight-song." It is the language of a man who cares.

And what of the semipalmated sandpiper, a few of which I last saw at low tide on Labor Day? Is it appropriate to use words such as gossip and love, to think of their self-awareness? I put the question to the British ornithologist Tim Birkhead, whose latest book is *Bird Sense: What It's Like to Be a Bird*. He told me he couldn't recall any behavioral tests of sandpipers, nor rigorous comparisons to crows or parrots, but still, he said: "You can guess that they have more sophisticated cognitive abilities than most people would give them credit for." Given everything we know about animal consciousness, and the primal nature of both our own emotions and our social bonds, it certainly seems reasonable to err on the side of personalizing the birds.

Birkhead told me an anecdote about a red knot—*Calidris canutus*, a close relative of the semipalmated sandpiper—found injured in 1980 on the north Dutch coast by a middle-aged couple. Jaap and Map Brasser named him Peter and nursed him to health. Peter never flew again, and lived with the Brassers and their dog Bolletje for nearly twenty years. Each afternoon he received half a loaf of bread, not so much to eat as to peck; Peter felt an instinctive need to forage, and became agitated if he couldn't. At night he rested quietly at their feet, stirring when wildlife shows came on television. He and the dog became companions. Years after Bolletje died, recordings of his barks brought Peter running.

That Peter would bond with a dog isn't so unusual. Red knots are social birds and, as we've seen, sociality is a great evolutionary driver of consciousness. What was unusual was a change in Peter's internal clock, which naturally guided his migratory transformation. Rather than following the seasons, it became synchronized to his new family. The ornithologist Theunis Piersma, as recounted by Dutch nature writer Koos van Zomeren, speculated that Peter "developed his own personal cycle and . . . stayed red as long as possible hoping that Jaap, Map, and the dog would also become fat and change colour, after which they would all depart for Greenland."

Of course, the Brassers knew Peter well, whereas I've only glimpsed semipalmated sandpipers. I can't truly know what goes on their heads. Yet at some point this becomes irrelevant: we can't ever really know what goes on in another person's mind, but we manage all the same. I'm happy to know simply that the birds I've seen have their own private worlds, their own sense of light and companionship. They go to sleep expecting to wake again. Perhaps they have names for each other. I just don't know what they are.

MONOGAMY HELPS GEESE REDUCE STRESS

With monogamy so uncommon in the animal world, the idea of lifetime fidelity can seem a little strange, at least to evolutionary biologists.

But in greylag geese, who can live for twenty years and share those years with just one mate, biologists have found a benefit: stress reduction. During fights, males with mates have lower heart rates than their single brethren. If their partners are nearby, they're even more relaxed.

"It seems to be one advantage of monogamy: If you have a long-term relationship, that helps you cope with stress. It's a good thing to do compared to systems where you always have to find a new partner and have no support in such situations," said Claudia Wascher, an ethologist at Austria's Konrad Lorenz Research Station.

For their study, published in *Biology Letters*, Wascher's team looked at data taken from heart rate monitors implanted in twenty-five greylag geese.

In earlier research, she'd observed how greylag heart rates reflect the birds' stress levels and can fluctuate wildly during confrontations, which occur often among these highly social, close-living birds.

But Wascher had noticed that whereas some geese became extremely stressed during fights, others stayed relatively calm. In the new study, she looked for an explanation; it proved to be the geese's mate status. Average heart rates of males with partners were about 10 percent lower during fights than the heart rates of single males.

While resting, female heart rates rose when their mates were more than a few feet away. Stress dissipated when they returned.

"An increase in heart rate is an investment. You'll have more energy available. But if you don't do that, if you're in safe mode, you'll save more energy, which could be a long-term advantage," Wascher said.

Among humans, multiple studies show links between partnership and responses to stress and disease.

Whether and how partnership-induced physiological changes are subjectively experienced by the geese is an open question. Konrad Lorenz, the Nobel-winning ethologist for whom Wascher's research institute is named, famously believed that greylag geese felt love.

Wascher is reluctant to describe the birds' feelings, but noted that when a mate dies a goose's heart rate will drop precipitously and stay low for up to a year. Bereaved geese will eventually find new mates, but for a long time they appear to be sad.

"There is a long-term physiological change, a decrease in metabolic rate and function. You can find terms like depression-like syndrome in the literature, though this is something that is hard to say," she said. "We are very careful with these analogies. But as intelligent as humans might be, we're still determined by our physiology. Everybody knows that having a social partner in humans is an important thing. And we're all socially living vertebrates."

Not all greylag geese partnerships are heterosexual. Male-male pairs are not uncommon, and triads with two males and one female have been observed. In Wascher's flock, there were two homosexual pairs.

"We didn't find any difference between male-male and male-female partners," said Wascher. "It could be because there were only two pairs, but my guess would be that there's not really a difference."

WHAT PIGEONS TEACH US ABOUT LOVE

Last spring I came to know a pair of pigeons. I'd been putting out neighborly sunflower seeds for them and my local Brooklyn house sparrows; typically I left them undisturbed while feeding, but every so often I'd want to water my plants or lie in the sun. This would scatter the flock—all, that is, except for these two.

One, presumably male, was a strapping specimen of pigeonhood, big and crisp-feathered in an amiably martial way. The other, smaller bird presented a stark contrast: head and neck feathers in patchy disarray, eyes watery, exuding a sense of illness that transcends several hundred million years of divergent evolution.

She didn't have the energy to take wing as I approached. Instead she'd take several desultory steps away. Her mate would fly to the deck railing, where he paced back and forth. He gave every impression of wanting to flee—but not without his mate, at whom he looked back with apparent concern. This caught me by surprise. I spend a fair amount of time watching animals and writing about them—not just about their populations or interactions or physiologies, but about their minds, what they might think or feel—yet I hadn't much tried to put myself in a pigeon's feathers, so to speak.

Moreover, I slipped into that easy habit of interpreting behaviors through a narrowly evolutionary lens, assuming their decisions to be coldly calculated to maximize reproductive success. From which perspective the male's loyalty made little sense: Better for him to fly off and find another, healthier mate with whom to pass on his genes than to stick around with this sick bird.

Of course, I wouldn't frame my own life that way. Where I have meaningful feelings, they would have imperatives. Yet as I watched Harold and Maude, as I so unoriginally named them, their drama unfolding beside murals my girlfriend and I had painted as expressions of our own feelings, I began to wonder. Harold behaved in a manner expressive of devotion, tenderness, and affection: the foundations of what in humans we call love.

It's a word not often associated with pigeons, or even other animals. "Our highest esteem is accorded romantic love, which is considered the most suspect to ascribe to animals," writes Jeffrey Moussaief Masson in *When Elephants Weep: The Emotional Lives of Animals*. Indeed, science for most of the last several centuries would have found the suggestion risible, suggesting instead that Harold felt—if pigeons could even be said to feel—some instinctive, unconscious urge to stay nearby, an urge with no more emotional resonance than an itch.

Love, after all, is central to the human condition. How could a creature with a brain the size of a bean possibly feel something so profound? Something that gave rise to *Romeo and Juliet* and "Unchained Melody" and the Taj Mahal?

Part of the reluctance to talk of bird love, I suspect, is rooted in our misgivings about our own love's biological underpinnings: Is it just chemicals? A set of hormonal and cognitive patterns shaped by evolution to reward behaviors that result in optimal mating strategies? Perhaps love is not what defines us as human but is something we happen to share with other species, including the humble pigeon.

City dwellers often see pigeons as an eyesore and a nuisance. The more nature-inclined regard them as marvels of natural history and urban adaptation. Descended from birds bred by European hobbyists, *Columba livia* now nest on building ledges rather than their ancestral cliffs. They scratch out sustenance from refuse, handouts, and the seeds of weeds, and have become symbols of a certain indomitability.

But can pigeons be in love? Considering the possibility, it's worth stepping back and looking at where society and its knowledge-defining practices now stand in regard to the notion of nonhuman thoughts and feelings.

The old habit of treating other animals as so many furred and feathered automata—Descartes famously likened animals to clocks—is in fast recession. Scientists talk regularly about animal intelligence. But that automatic habit shaped scientific discourse and public imagination. Every assertion of complex experience could be met by the default rebuttal of anthropomorphism: Might we merely be projecting human qualities onto something much simpler, even alien?

Its legacy is still felt. Animal consciousness tends to be most appreciated in a select class of animals: big-brained creatures like great apes or whales, or domestic companions like cats and dogs, who can't be ignored. As a class, birds receive comparatively little attention. And when they do, it tends to focus on intelligence, on easily quantifiable feats of problem-solving and cognition, rather than emotion. Most anyone who follows science knows about brainy crows using tools and high-level reasoning. But avian love remains beyond the pale.

A telling example is *Partnerships in Birds: The Study of Monogamy,* a collection of studies published in 1996 with the express intent of explaining why birds are monogamous, which makes not a single mention of emotion. Affection appears once, in connection to a brief mention of attachment in so-called pair bonds between mates; attachment, readers are reminded, need not be understood in potentially loaded terms of strength or weakness, but rather measurements of "proximity and synchrony of behaviours which may influence fitness." It's a fascinating book, but also slightly ridiculous, like watching old video of lawn tennis matches, in which custom dictates the players wear white slacks and not run too hard.

The conservatism is understandable: Feelings are hard enough to measure in humans, much less animals, and "you can't think of birds as little humans," says Kevin McGowan, a Cornell University ornithologist who specializes in the social behavior of crows. Yet evolution is conservative too, notes McGowan, shaping the animal kingdom's diversity from common biological elements. Of emotions, McGowan says, "there's no reason to think that we humans have some brand-new thing."

Indeed, love's essential biology is evolutionarily ancient. Oxytocin and vasopressin, the hormones most closely associated with mammalian bonding, have the near-identical avian analogues of mesotocin and vasotocin, which shape the interactions of zebra finch couples. Birds likewise possess the basic reward-system neurotransmitters serotonin and dopamine. Birds might not have much in the way of easily recognizable facial expressions, but their biochemistry's symphonic chain reactions play out in neurological structures that evolved early in life's history, long before the cerebral cortex itself.

That alone is no guarantee of romance. Jane Goodall, the legendary primatologist who so powerfully described the abiding love of chimpanzee mothers for their children, has written that she cannot conceive of our closest living relatives as experiencing anything comparable to romantic love. To Goodall, chimp courtship is too brief to permit deep feelings. Their proclivities, she has noted, were not shaped by evolutionary circumstances conducive to love, namely, long-term relationships with single partners, which are the norm for modern humans.

In this respect we diverge markedly from chimpanzees—but not from birds, in whom monogamy is found in some 90 percent of all species, including pigeons. Neither their nor our monogamy is a pure, idealized sort, exclusive of infidelity or a succession of partners. Extra-pair copulation, or what we call cheating, can be quite common. But monogamy is the baseline and pigeons, who frequently mate for life, are among the more fidelitous of birds. Within the evolutionary context of monogamy, a capacity for love makes perfect sense.

Monogamous couples share food, information, and offspring-rearing duties, especially in species whose young are born requiring constant care—as is the case with pigeons, their helpless chicks so carefully hidden that few city-dwellers have ever seen one. Love—an attentiveness to the needs of another being, reinforced by emotional rewards—should enhance cooperation and improve a couple's chances of raising healthy offspring. And as noted by Claudia Wascher, a zoologist at

Anglia Ruskin University, whose PhD research described how mated greylag geese have lower levels of stress hormones than single birds, there's no question that pair bonds are powerful.

"Social bonds in general seem to be terribly important for birds," says Wascher, "and the most important social bond for most birds is the pair bond." Monogamy, then, should be fertile evolutionary ground for love's blooms.

McGowan and Wascher readily recognize emotion in birds. "I would suspect they do have affection for each other," says McGowan, who has observed crow couples stay together for more than a decade. "It's not going to be the same as what humans have, but I suspect it's close enough that we'd recognize it," he says. Yet McGowan stops short of love: Science describes behaviors easily, but is in murkier terrain with complex states of mind.

Indeed, it might seem a flight of fancy to equate pair-bonding's neurobiological rewards with love, no matter how much evolutionary sense it makes. Pigeons have the necessary pieces and life history, but can their experiences of bonding really compare to what in humans inspired the eighth-century poet Chang Chi's lamentation of the unrequited: "So I must give you back your pearls / with two tears to match them"? Can pigeon-brained attachment manifest in love's full spectrum, from butterflies-in-the-stomach infatuation to the ecstasy of consummation?

It's still possible, however, to imagine that avian love is more than a mindless itch. Perhaps human love is unusually complex, invoking not just physiology but our unique cognitive sophistication. Still, many species display a cognitive complexity—awareness of self and others, long-term memory, a capacity for abstract concepts—comparable to primates. The gentle social courtship of "allopreening," in which birds groom each other's feathers, is especially sophisticated. Just as I can think fondly of my lover while she's away, so might a pigeon think fondly of its absent mate.

We can consider observational evidence to buttress the biological. About a decade ago, Rita McMahon found a pigeon with a broken leg on her deck in New York City's Upper West Side. The pigeon was otherwise quite fortunate. McMahon would go on to cofound the Wild Bird Fund, which cares for some 3,500 sick and injured birds every year. A veterinarian amputated the pigeon's leg; while it recovered, it would rest on a cushion in McMahon's apartment window. On the other side stood her mate, day after day, keeping her company until she was released and the couple rejoined.

"They were devoted to each other," says McMahon, who also recalled how one of her volunteers once found a broken-winged robin in a depression in a snow bank, his mate nearby. The volunteer picked up the injured bird and put him in a bag for transport to the hospital. With little fuss she then gathered the mate—which was quite unusual, as healthy wild birds are uniformly skittish. "I understand being able to pick up a broken-winged robin easily, but not one who's intact," McMahon says. At the hospital, they learned that the break wasn't fresh. The robin was in surprisingly good health. His mate, believes MacMahon, had been taking food to him on the snowbank, "and decided to stay with her man."

Love is as love does. "There's no reason to think it would be much different for humans than nonhumans," says Marc Bekoff, author of *The Emotional Lives of Animals*. "I've known mourning doves"—a species closely related to pigeons—"who were more in love than a lot of the people I've known." To Bekoff, love's ultimate measure is the presence of its converse, grief.

Apparent grieving exists in the avian world, most notably among greylag geese, in whom individuals who've lost a partner display the classic symptoms of human depression: listlessness, a loss of appetite, lethargy lasting for weeks or even months. The same applies to pigeons. On Pigeon Talk, a website of pigeon-breeding hobbyists, anecdotes abound of birds sinking into a funk after losing their mates, and sometimes refusing to take another mate for up to a year afterward—no small time for a species that typically lives for less than a decade.

One of the most moving stories involves mourning doves. After a dove was eaten by a hawk in the backyard of a forum member called TheSnipes, the mate stood beside the body for weeks. "I finally couldn't stand to watch it any more and picked up every feather and trace of remains that was left there and got rid of it," wrote TheSnipes. "The mate continued to keep a vigil at that spot though, for many months, all through the spring and summer."

McMahon noted something I hadn't considered: There are good and bad pigeon couples. Some are attentive and physically affectionate, constantly stroking one another's feathers. Others appear distant and peckish. As human love varies, so might theirs. Not every pigeon's tale need be so romantic as *Fly High, Fly Low*, Don Freeman's delightful children's book about the search of Sid for his mate Midge, lost to him—though only for a while—when workers take down the sign in which they've made their nest. Others might better resemble Maud and Claud of Patricia Highsmith's "Two Disagreeable Pigeons," regarding each other with pique and scorn, kept together by inertia and habit.

It's also worth considering whether pigeons might experience aspects of love that we don't. Could a bird whose basic physiology adapts to changing seasons, who can perceive atmospheric infrasound and see Earth's magnetic field, have emotional capacities beyond our own? Including, perhaps, forms of love that are not merely analogues of our cherished feelings, but something unique to them?

It's something to imagine. "Love among animals might appear as mysterious and baffling as human love has over the centuries," writes Masson. At risk of sounding unromantic, though, I'm not convinced that love is so mysterious. It just feels good.

As for Harold and Maude, I don't know how their story ended, or indeed whether it continues. They roosted in a partially abandoned building on my fast-gentrifying block. It's now being turned into condos, making them victims of Brooklyn's rising real estate prices, albeit with a better chance than most humans at finding a decent place to live nearby.

Their example stayed with me, though, and now colors the way I think of my winged neighbors. Ubiquitous and unappreciated, typically ignored or regarded as dirty, annoying pests, pigeons mean something else to me now. Perched on building ledges, chasing scraps of food, taking to the skies at sunset: each one is a reminder that love is all around us.

CHIMPS AND THE ZEN OF FALLING WATER

There is a waterfall in Tanzania's Gombe National Park. Maybe 12 feet high, it's fairly modestly sized, though even a modest waterfall is quite a magical thing. And it's here that chimpanzees come to dance.

You can watch a video online, narrated by the great primatologist Jane Goodall, who, as with so many chimpanzee behaviors, was the first to observe these rituals. It's quite a show: An adult male approaches via the riverbed with a slow, rhythmic gait, so unlike yet like our own. He throws rocks and tree branches into the falls, then catches a vine and swings above them. Finally he sits on a rock in the stream, head resting on forearms, and watches the water go by.

These performances happen several times each year, but they do not have any obvious utilitarian purpose, even as social displays; though groups of chimps sometimes participate, often it's just a solitary individual. Given how easily a chimp might slip on the rocks, it's quite risky. Chimps also can't swim, and typically avoid running water. Which raises the question: What are these chimps thinking?

"I can't help feeling that this waterfall display, or dance, is perhaps triggered by feelings of awe and wonder," says Goodall in the video. "The chimpanzee brain is similar to ours. They have emotions that are clearly similar to those that we call happiness and sadness and fear and despair and so forth. So why wouldn't they also have feelings of some kind at spirituality? Which is, really, being amazed at things outside yourself."

In which case: *spiritual chimps.*

It's been a hard notion for some to accept. The theologian Christopher Fisher argued that Goodall was anthropomorphizing her subjects—interpreting an animal's behavior in inappropriately human-centered terms—and that only humans can be spiritual. Although Fisher's refutation could just as easily be dismissed as yet another example of anthropocentrism, assuming that what's important to humans is also what's singular about us, he raises a fair point: We don't know what's happening in a rain-dancing chimp's mind.

Then again, if you want to get philosophical about it, we can't really know what's going on in each other's minds, either. We just weigh evidence and make inferences and put together some subjective but educated guesses. And this we do know about chimps: They possess the cognitive capacities—the ability to think symbolically and use imagination, and to learn through imitation—that underlie human ritual and spirituality.

"Perhaps it is a ritual of respect or abnegation to the god of water," writes the primatologist William McGrew in *The Cultured Chimpanzee*. This is, he cautions, a speculation that can't be tested. It's also possible that the chimps use the waterfall and its predictable stimulus—"Can anyone not stand at the base of a waterfall without their pulse racing?" asks McGrew—as a setting to practice displays of bravado typically expressed in confrontations with other chimps. Sparring practice, essentially.

Mary Lee Jensvold, a primatologist and former director of the Chimpanzee and Human Communication Institute, offered her own thoughts. Like McGrew, she posited an alternative explanation: various loud stimuli, including machinery and boisterous people, can elicit chimpanzee displays. Yet what "makes me think this is something deeper," says Jensvold, "is the sitting quietly and staring at the waterfall afterwards."

That's what gets me, too. A chimp just sitting with his or her thoughts. Contemplating the water, much in the manner I do at streamside. I know where the water comes from, the planetary cycles and local ecologies, and still it's a magical current of time and life. Sometimes I say a thank-you. What might water evoke in a being capable of rich abstraction, eminently aware of water's vitality, but with a different and less mechanistic body of knowledge? Suddenly it doesn't seem so far-fetched to imagine chimps experiencing some seed of what, in *Homo sapiens*, eventually grew into water myths found in just about every human society. Spirituality in its elemental form is pretty simple. Like Goodall says, it's being amazed at things outside yourself.

"I think chimpanzees are as spiritual as we are, but they can't analyze it," Goodall says in the video. "You get the feeling that it's all locked up inside them, and the only way they can express it is through this fantastic rhythmic dance." Maybe these waterfall dances—and also similar dances she's observed at the start of sudden, seasonal downpours—will someday give rise to animistic proto-religion, imbuing falling water with existential meaning. Perhaps they already have.

What about other nonhuman primates? I asked the UCLA anthropologist Susan Perry, who studies capuchin monkeys. Sometimes they perform dances otherwise conducted during courtship when the first rains come after dry season, she said. It doesn't happen every year, though, and Perry doesn't think it displays spiritual wonder or awe. In a similar vein, the anthropologist Laurie Santos of Yale University reports nothing comparable from her rhesus monkey observations.

This would seem to fit with what's known of monkey cognition: They do think in complex ways, and certainly possess some elaborate interpersonal rituals, but they're not quite so abstract-minded as chimps. Outside of primates, though, plenty of species have the mental potential. Quite a few cetaceans, for example, including orcas with their remarkable tribal greeting ceremonies. Ditto elephants and their burial rites. Again: we can't know what they're thinking, but it's unscientific not to consider the possibility. "Perhaps numerous animals engage in these rituals," writes the ethologist Marc Bekoff in *The Emotional Lives of Animals*, "but we haven't been lucky enough to see them."

And what else might be going on with chimpanzees? One particular report, published several years ago in the *American Journal of Physical Anthropology*, catches my imagination. In it researchers describe a group of chimps living at Fongoli, Senegal, in a savanna reminiscent of settings where the earliest humans evolved. The chimps, wrote the researchers, often dance at the edge of fires.

HOW CITY LIVING IS RESHAPING THE BRAINS AND BEHAVIOR OF URBAN ANIMALS

When next you meet a rat or raccoon on the streets of your city, or see a starling or sparrow on a suburban lawn, take a moment to ask: Where did they come from, so to speak? And where are they going?

In evolutionary terms, the urban environments we take for granted represent radical ecological upheavals, the sort of massive changes that for most of Earth's history have played out over geological time, not a few hundred years.

Houses, roads, landscaping, and the vast, dense populations of hairless bipedal apes are responsible for it: all this is new. A growing body of scientific evidence suggests that the brains and behaviors of urban animals are changing rapidly in response.

"A lot of biologists are really interested in how animals are going to deal with changes in their environments," said the biologist Emilie Snell-Rood of the University of Minnesota. "Humans are creating all these totally new environments compared to what they've seen in evolutionary history."

Snell-Rood is one of many researchers who have updated the conventional narrative of urban animals, in which city life favors a few tough, adaptable jack-of-all-trades—hello, crows!—and those species fortunate enough to have found a built environment similar to their native niches, such as the formerly cliff-dwelling rock doves we now call pigeons and find perched on building ledges everywhere.

The long view, though, is rather more multidimensional. Cities are just one more setting for evolution, a new set of selection pressures. Those adaptable early immigrants, and other species that once avoided cities but are slowly moving in, are changing fast.

As Snell-Rood and colleagues describe in a *Proceedings of the Royal Society B* article, museum specimens gathered across the twentieth century show that Minnesota's urbanized small mammals—shrews and voles, bats and squirrels, mice and gophers—experienced a jump in brain size compared with rural mammals.

Snell-Rood thinks this might reflect the cognitive demands of adjusting to changing food sources, threats, and landscapes. "Being highly cognitive might give some animals a push, so they can deal with these new environments," she said.

Brain size is, to be sure, a very rough metric, one that's been discredited as a measure of raw intelligence in humans. For it to fluctuate across a whole suite of species, though, especially when other parts of their anatomy didn't change, at least hints that something cognitive was going on.

Many other studies have looked at behavior rather than raw cranial capacity. In these, a common theme of emerges: Urban animals tend to be bold, not backing down from threats that would send their country counterparts into retreat. Yet even as they're bold in certain situations, urban animals are often quite wary in others, especially when confronted with something they haven't seen before.

"Maybe avoiding danger is a useful trait for some animals living in urban environments," said the biologist Catarina Miranda of Germany's Max Planck Institute, who in a *Global Change Biology* paper described her experiments with rural and urban blackbirds.

"Most of the birds that never approach new objects or enter new environments in this long period of time are urban," Miranda said. "There are many new dangers in a town for a bird. Cars can run you over. Cats can eat you. Kids can take you home."

Somewhat counterintuitively, bold urban animals also tend to be less-than-typically aggressive, a pattern documented in species as disparate as house sparrows and salamanders, the latter of which are a specialty of Jason Munshi-South, an evolutionary biologist at the City

University of New York. The city's salamanders—there aren't many, but they're there—"tend to be languid," said Munshi-South. "If you try to pick them up, they don't try to escape as vigorously as they do outside the city. I wonder if there's been natural selection for that."

If so, it might be driven by high population densities of salamanders in the city. Aggressive neighbors don't tend to be good neighbors. Through that lens, city animals could be domesticating themselves, a process that can occur without direct human intervention.

Even more fundamentally, muted stress responses have been found in many species of urban animals. When surprised or threatened, their endocrine systems release lower-than-usual amounts of stress hormones. It's a sensible-seeming adaptation. A rat that gets anxious every time a subway train rolls past won't be very successful.

"They're clearly attenuating their physiological response to stress, probably because they're constantly inundated with noise, traffic, and all kinds of environmental stresses in cities," said the biologist Jonathan Atwell of Indiana University. "If they were ramping that response up all the time, it would be too costly."

A challenging question is whether traits like these represent inherited biological changes or what researchers call phenotypic plasticity: the ability to make on-the-fly adjustments to circumstance.

Some adaptations, such as the swath of genetic mutations that Munshi-South identified in New York City's white-footed mice, are clearly heritable. Others are learned. In many cases, both processes are likely involved, said Atwell, who studied the question in his research on songbirds called dark-eyed juncos around San Diego.

The San Diego juncos sing at higher frequencies than those living in rural, traffic-free settings. When Atwell raised some of their chicks in a quiet place, that rise in song frequency dropped by about half, suggesting an even split between heritability and plasticity.

Where things get really interesting, though, is with social learning and animal culture—all those animal habits and abilities that are not inborn, but taught. "I suspect that often it's not their cognitive abilities evolving, but cultural evolution going on," said Atwell. "Anytime animals can learn behavior from one another, I think there might be cultural evolution."

Urban squirrels, for example, seem to have adjusted to vocalization-drowning ambient noise by making tail-waving a routine part of communications. Perhaps this was instinctive in a few animals, then picked up by others. Likewise, squirrels might learn about traffic by seeing others get run over, said Snell-Rood. Rats could see brethren die after eating poisoned bait, then teach pups to avoid the traps.

These possibilities are only hypothetical, but hardly implausible. After all, other animals traditionally recognized as clever—such as crows who share information about untrustworthy humans, or temple-dwelling Asian monkeys who pickpocket tourists—are clearly learning about us, and intelligence has been studied only in a few species.

Not all changes in urban animals will represent adaptations to urban living, however. Most genetic mutations are neither beneficial nor harmful, at least not right away. They simply happen and, over long periods of time, accumulate in populations through what's known as genetic drift.

In isolated groups, drift's effects are magnified, as are so-called founder effects, in which entire populations bear the genetic imprint of a few early animals. For these creatures, urban adaptations won't necessarily represent adjustments to city life, but rather simple happenstance.

How might this play out in deep time? If humans can keep civilization intact long enough, will urban animal populations eventually become their own distinct species—bold, relaxed, and clever, with a store of learned information about our habits, and perhaps a few other traits that arise by chance?

Nobody knows, said Snell-Rood, but "you could imagine some kind of speciation over long periods of time." She noted, though, that not all the changes seen in urban animals are necessarily permanent. The big brains of those city-dwelling Minnesota mammals, for instance, seemed to shrink after a few decades of urban adaptation.

"The way I interpret it is that during the initial colonization, it pays to be smart," Snell-Rood said. Once city life becomes predictable, "you can go back to having a smaller brain."

RECONSIDER THE RAT: THE NEW SCIENCE OF A REVILED RODENT

Many people live among animals that inspire affection, even admiration: bluebirds and hawks, beavers and bobcats. The fortunate might glimpse a fox or, if they live in the country, an eagle or a bear. But I live in New York City, where the star of the show is the rat.

Pity the rat. Few mammals are so reviled, so deeply and viscerally loathed in Western culture as *Rattus norvegicus*. Regarded less as animals than as rapacious, beady-eyed vectors of filth and disease, they can drive otherwise warm-hearted souls to fantasies of extermination. Even rats on an abandoned ship, to take a recent example, aren't seen as starving; they're reviled as cannibals.

Yet even those who recoil at the sight of a hairless tail often grant rats a certain respect. Despite centuries of ruthless extermination, *R. norvegicus* has withstood everything we've thrown at it and come back for more. Above all, they are survivors. A rat foraging on a subway track, thriving on what we discard and overlook, has a certain outlaw charm.

Often I've found myself considering what scientific questions might be asked of these supposedly lowly creatures. Plenty of people ponder consciousness in whales, or wonder what life looks like to a chimpanzee or a cat. But what about rats?

Unexpected as it might seem, we still have much to learn about rats, and from them. Yes, there's volume upon volume of rat research—but most of it focuses on traditional questions of basic physiology and drug responses and so forth. Few researchers have asked what rats think and feel, or how they've adapted to environments so very different from their ancestral home in southern Mongolia.

On this front, rats are guides to emerging questions of evolution and cognition: how cities shape the brains and behaviors of animals within them, and whether aspects of consciousness once considered exceptional might in fact be quite common.

Foremost among these is empathy, widely considered a defining human characteristic. Yet rats may possess it too. An especially fascinating line of research, the latest installment of which was published last year in the journal *eLife*, suggests rats treat each other in an empathic manner. Such thoughtfulness underscores the possibility that rats are far more complicated than we're accustomed to thinking—and that much of what's considered sophisticated human behavior may in fact be quite simple.

This idea runs contrary to notions of human exceptionality. Yet evolution teaches us that humans and other creatures share not only bodies but brains as well. In that light, why wouldn't rats care about one another? The idea also challenges us to see rats anew: not just as vermin, or as anonymous laboratory models of some biological process, but as fellow animals.

As the neurobiologist Peggy Mason, a pioneer in rat empathy research, put it, "I'm perfectly happy thinking of myself as a rat with a fancy neocortex."

On a table in Mason's University of Chicago lab sits a plexiglass box about two feet square. Inside is a white Sprague-Dawley rat, a strain bred for laboratory study, and a plexiglass canister holding a black-and-white Long-Evans rat.

The trapped Long-Evans is clearly agitated. The white rat is too. Instinctively, she wants to stay in the corner; rats avoid open spaces and navigate by touch, which is why you often see them scurrying along walls. Yet she rushes again and again to the canister, sniffing at the rat inside, nosing the glass, nudging the door. Eventually, she opens it, freeing the rat. They rub together.

At a purely descriptive level, you could say one rat helped another. Why that happened is the question. According to Mason and her collaborator Inbal Ben-Ami Bartal, the free rat appears to empathize with her trapped comrade. She recognized the rat's distress, grew distressed herself, and wanted to help. This appears to be a powerful impulse in rats. In tests of whether rats would rather eat than help another rat, the researchers found empathy's pull to be as strong as their desire for chocolate—and rats do love their chocolate.

The two researchers first claimed rats might feel empathy in a high-profile 2011 *Science* paper describing rats freeing their cagemates. They expand on those findings in the latest study, which describes rats helping strangers. It's a radical, even controversial, claim. Some scientists recognize that chimpanzees, a few cetaceans, and perhaps elephants could be empathic, but few have ascribed that trait to rats. If *R. norvegicus* can be empathic, that fundamentally "human" trait might in fact be ubiquitous.

"We're in a period of transition with respect to how we think about animals," said the environmental philosopher Eileen Crist. After centuries of seeing the animal kingdom as a hierarchy with humans on top, of treating animals as purely instinct-driven biological machines, "cognitive ethology is opening up a new terrain. Knowledge itself is fluid and changing right now"—and empathy investigations are very much a part of that.

Those who've had pet rats may not be surprised by reports of their empathy, nor will readers of naturalists' texts from the nineteenth and early twentieth centuries. (Witmer Stone and William Everett Cram, for example, wrote of rats in 1902's *American Animals*, "Careful witnesses have always given them credit for looking after any helpless member of their family.") But informal observations carry little scientific weight, and researchers are reluctant to describe what animals might think and feel. After all, animals can't tell us, and we can't read their minds.

There's some historical baggage, too. Twentieth-century study of animal behavior was famously inhospitable to the idea that animals feel much of anything. B. F. Skinner, the father of modern animal behavioral science, called emotions "an excellent example of the fictional causes to which we commonly attribute behavior." Such views have largely fallen from favor, but science has been slow to embrace Charles Darwin's essential point: that humans and other animals necessarily share not only anatomical roots but neurological origins as well.

Claiming empathy for rats isn't easy, and one criticism of Mason and Bartal's interpretation is that a far simpler phenomenon called emotional contagion could explain their rats' helpfulness. In other words, when one rat becomes distressed, that distress spreads to others—but they don't necessarily feel for the first and translate that feeling into intention.

As the Oxford University zoologist Alex Kacelnik and colleagues noted in a 2012 *Biology Letters* reflection on empathy research, some ants display helping behaviors similar to Mason and Bartal's rats. "Any solid evidence for empathy in non-humans would be a notable advance," they wrote, "but, in our view, it remains unproven outside humans."

Other researchers defended the possibility of rat empathy. "Ants are not rats," quipped Frans de Waal, an Emory University ethologist who has written extensively about empathy, on Facebook. "It would be totally surprising, from a Darwinian perspective, if humans had empathy and other mammals totally lacked it." As for Mason and Bartal, they've downplayed the empathy interpretation in their latest work, restricting it to speculative discussion.

In those experiments, they observed that rats helped not only their cagemates but total strangers as well. They even helped strangers from other strains, like the Sprague-Dawley helping the Long-Evans, if they'd previously known a rat from that strain. Social experience mattered more than narrow biological self-interest. Regardless of whether that's empathy, Mason and Bartal hope the implications could extend beyond rats and provide a model system for investigating the basic biology of helpfulness.

"It's our fervent hope that this model will be used by many researchers to look at helping behavior," Mason said. "If we can see a way that rats can form an affective bond quickly, can we use that in society? Wouldn't that be useful?"

Although she and Bartal downplay the possible role of empathy, they still think it's the most likely explanation for what they've observed. This argument isn't just based on their experiments, but on evolutionary neurobiology. The brains of humans and rats are certainly not identical, but they overlap in fundamental ways. "We share the same neural structure with rats that we use for our own empathic responses," said Bartal.

The thing to remember, the researchers say, is that empathy can take different forms. We often focus on its most sophisticated definition: the empathy that comes with, say, reading about the suffering of people far away and feeling compassion for them. So abstract a perspective is almost surely unique to *Homo sapiens*. But empathy can be much simpler, as when we instinctively want to comfort someone who's crying. Most of what we consider human empathy, said Mason, takes this form, and needn't be unique to us. It could exist in various permutations across mammals, as do its underlying evolutionary drivers and neurobiological processes.

Many well-regarded psychologists and neuroscientists have taken this position in recent years, arguing that simple empathy provides obvious evolutionary benefits for social animals, especially those species in which mothers care extensively for their young. Even complex, higher-order human empathy appears to stem from basic emotional and cognitive processes that rats—indeed, all mammals—certainly possess. "Evidence is accumulating that this mechanism is phylogenetically ancient, probably as old as mammals and birds," de Waal wrote in a 2008 *Annual Review of Psychology* paper.

Jaak Panksepp, a Washington State University neuroscientist renowned for his research on rat emotions, says these lines of research raise a fascinating question: How are empathy's various forms driven by simple mental processes, and to what extent do they involve complex, high-level cognition? If some forms of empathy can be quite simple, Panksepp says, the same could well apply to many other abilities long considered the sole province of humans.

One such ability, metacognition—the ability to think about thinking—already has been demonstrated in rats; as with empathy, simpler explanations could suffice, but the possibility is certainly there. Depending on the sorts of memory and reasoning involved, rats also may be able to think symbolically; build mental models of their physical and social worlds; imagine themselves in the past and future; and communicate in a sort of protolanguage, much like their prairie dog cousins appear to do. Such abilities would not be exactly like our own, but could resemble them far more than we generally think.

"Over the last few decades, comparative cognitive research has focused on the pinnacles of mental evolution, asking all-or-nothing questions such as which animals (if any) possess a theory of mind, culture, linguistic abilities, future planning, and so on," wrote de Waal. "A dramatic change in focus now seems to be under way, however, with increased appreciation that the basic building blocks of cognition might be shared across a wide range of species."

"We don't have good data to adjudicate those issues, but I'd agree those are very important ones," said Panksepp. "There should be an open conversation." Mason puts it more bluntly: "There's just no intellectual reason why we don't share a basic underlying behavioral fabric."

To get a sense of how these questions might be playing out in the wild, I spent some time with Jason Munshi-South, a biologist at Fordham University. A self-described specialist in "evolution in the Anthropocene," the informal scientific name for an epoch profoundly shaped by human activity, he's among the few scientists studying how city rats are evolving.

To better gather the tissue samples he uses to create population-genetic narratives of ecology and evolution, Munshi-South attended the exquisitely named NYC Rodent Academy. Taught by Robert Corrigan, who is generally regarded as the country's foremost expert on rat control, the class opened a new dimension of the city. "It was like learning how to read the landscape," Munshi-South said. "Everywhere I go, I see rat sign."

We start our safari in an alley in lower Manhattan, not far from City Hall. We see no rats—it's too cold, and there's too much activity at an adjacent construction site—but their signs are everywhere.

Munshi-South points out cracks between bricks and in foundations, barely noticeable yet unmistakably gnawed at their edge, entrances to tunnels leading to dens underfoot or behind walls.

We also see several of New York's ubiquitous and, in the grand scheme of things, mostly pointless poison-bait boxes. Rat control is really food control, teaches Corrigan: so long as they have plenty to eat, rats will flourish. Cleaning up is hard to do, though, so instead we kill. Around one box are scattered nibbled pink bait blocks. They contain grain and potent blood thinners, and the rats who ate them probably bled to death, or soon would.

Or maybe not. Many rat populations are becoming resistant to blood thinners, just as they adapted to previous rodenticides. Munshi-South wonders what specific resistance mutations have arisen within New York City's rats and how populations respond to eradication efforts. Rats in each neighborhood might descend from a few hardy survivors, or maybe new generations move in after previous residents die. These dynamics could affect how diseases ebb and flow in rats, and perhaps how humans are exposed to them.

Munshi-South suspects rats travel, like most New Yorkers, through subway tunnels, with genetic distributions displaying patterns of flow along the city's main train lines. We leave the alley and walk to another favorite trapping site, the ancient Chambers Street/Brooklyn Bridge-City Hall station. Again he notes the telltale cracks in walls and along tracks. It's easy to think they lead to an underworld rat metropolis, but of course they don't. Rats stay close to food, and there's plenty to be had here, a cornucopia of takeout boxes and illegally dumped trash.

Last summer, said Munshi-South, garbage bags piled at platform ends literally rippled with rats. Today they're still. A rodenticide caution sign indicates a recent round of poisoning. Two earlier dates on the sign have been crossed out, testament to the ongoing struggle. When I return a few days later, a rat forages along the tracks. "My guess is that in here, they'll never get rid of all of them," he said.

Do these rats survive simply because they become resistant to poisons, and for the mechanistic reasons often invoked when discussing evolution: dietary habits, reproductive rates, adaptation to life underground? Those certainly play a part, but perhaps not the only one. Mental and social evolution—rats growing smarter, and perhaps more empathetic—could be a factor.

"What I'd imagine with rats, moving into a new environment and heavily persecuted, is that there's selection on behavior that has a heritable basis, and especially on things like personalities," Munshi-South said. He plans to compare the genomes of city rats and lab rats, and look for genetic changes linked to cognition. "I don't think we know yet, but these are good hypotheses," said Munshi-South.

Neophobia, or a temperamental inclination to avoid new objects, is one possible area of adaptation. Curiosity kills the rat, so to speak. Spatial memory enhancements could be useful in a landscape more complex than the rats' ancestral plains. And empathy also could be key. One of its essential components, a tendency to pay attention to others and be aware of how they feel, is integral to social learning, or the ability to learn from others' examples.

"Given the importance of social learning in rats," said Emilie Snell-Rood, a biologist at the University of Minnesota, and its usefulness "in a situation where a novel predator like humans are trying to kill you all the time, I would expect increased selection on social learning." Empathy also could facilitate helpfulness, another potentially useful trait, and lead to diminished aggression: the more sensitive animals are to others' emotional states, the more socially agreeable they tend to be. That's been documented in animals living at high population densities, including chimpanzees, and urban rats are nothing if not populous.

Snell-Rood has studied how a century's exposure to humans changed Minnesota's small mammals. She hasn't examined rats, but did document significant increases in brain size among other urban rodents, including squirrels, shrews, and mice. That doesn't mean the animals are getting smarter, but it does suggest something cognitive is going on.

Whether that involves empathy is an open question, a hypothesis to be tested. It is speculation of the sort encouraged by a twenty-first-century appreciation of animal consciousness and evolution. Darwin would likely have enjoyed the possibilities. As he theorized in *The*

Descent of Man, "those communities which included the greatest number of the most sympathetic members would flourish best, and rear the greatest number of offspring."

In coming years, scientists could learn that *R. norvegicus*, which arrived in New York late in the eighteenth century, has taken an unexpected path to prosperity. Maybe city rats are kinder and gentler, at least with one another. It would make for a nice lesson. And it seems only appropriate that the animal making us reconsider empathy's nature and its role in evolution is a creature so universal, and so universally loathed, as the rat.

We probably won't leave out extra trash for them—but knowing that rats care for one another could help us appreciate them a bit more. "The basic response," said Panksepp, "is just wonder at the marvelousness of nature." We could even go a step further: in city landscapes and in our labs, we can try to imagine life from their perspective. We might even empathize with them.

MONKEYS SEE SELVES IN MIRROR, OPEN A BARREL OF QUESTIONS

Monkeys may possess cognitive abilities once thought unique to humans, raising questions about the nature of animal awareness and our ability to measure it.

In the lab of the University of Wisconsin neuroscientist Luis Populin, five rhesus macaques seem to recognize their own reflections in a mirror. Monkeys weren't supposed to do this.

"We thought these subjects didn't have this ability. The indications are that if you fail the mark test, you're not self-aware. This opens up a whole field of possibilities," Populin said.

Populin doesn't usually study monkey self-awareness. The macaques, described in a study published in the journal *PLOS ONE*, were originally part of his work on attention deficit disorder. But during that experiment, the study coauthor Abigail Rajala noticed the monkeys using mirrors to study themselves.

So-called mirror self-recognition is thought to indicate self-awareness, which is required to understand selfhood in others, and ultimately to be empathic. Researchers measure this with the "mark test." They paint or ink a mark on unconscious animals, then see if they use mirrors to discover the marks.

It was once thought that only humans could pass the mark test. Then chimpanzees did, followed by dolphins and elephants. These successes challenged the notions that humans were alone on one side of a cognitive divide. Many researchers think the notion of a "divide" is itself mistaken. Instead, they propose a gradual spectrum of cognitive powers, a spectrum crudely measured by mirrors.

Indeed, macaques—including those in Populin's study—have repeatedly failed the mark test. But after Rajala called attention to their strange behaviors, the researchers paid closer attention. The highly social monkeys only rarely tried to interact with the reflections. They used mirrors to study otherwise-hidden parts of their bodies, such as their genitals and the implants in their heads. Mark tests notwithstanding, they seemed quite self-aware.

"I think that these findings show that self-awareness is not an all-or-nothing phenomenon," said Lori Marino, an Emory University evolutionary neurobiologist who was not involved in the study. "There may be much more of a continuum in self-awareness than we thought before."

According to the Emory University primatologist Frans de Waal, the new findings fit with his work on capuchin monkeys who don't quite recognize themselves in mirrors, but don't treat the reflections as belonging to strangers. "As a result, we proposed a gradual scale of self awareness. The piece of intriguing information presented here may support this view," he said.

However, de Waal cautioned that "many scientists would want more tests and more controls"—a warning especially salient in light of a high-profile controversy involving Marc Hauser, a Harvard University evolutionary biologist who appears to have overstated the cognitive powers of his own monkeys.

"What you're seeing in the videos is subject to all kinds of interpretations," said Gordon Gallup, a State University of New York at Albany psychologist who invented the mirror test, and has administered it with negative results to rhesus monkeys. "I don't think these findings in any way demonstrate that rhesus monkeys are capable of recognizing themselves in mirrors."

Populin said his monkeys may have developed an unusual familiarity with mirrors, which are given to them as toys during infancy. The presence of saltshaker-sized implants screwed into their skulls may also have captured their interest more readily than an inked mark.

Marino, who helped demonstrate self-recognition in bottlenose dolphins, disagreed with Gallup. "The videos are absolutely convincing," she said. "I have been trying to find an alternative explanation for the results—and haven't come up with one yet."

Marino said the findings fit with other research on monkey cognition, including a since-replicated *Journal of Experimental Psychology* study in which macaques displayed unexpectedly sophisticated math skills and passed other, non-mirror-based tests of self-awareness.

"There are many ways to look at animals. Mirror tests are not the end-all and be-all," said Diana Reiss, a mammal cognition specialist at the City University of New York.

If research continues to find that monkeys possess higher-than-expected awareness, it could influence how researchers and the public think about biomedical research on monkeys. Macaques were critical in the development of a polio vaccine during the twentieth century and, more recently, the refinement of embryonic stem cell techniques.

"I would absolutely hope that we do not stop using them now. Their contributions have been immense," said Populin, who studies how Ritalin affects the brain's prefrontal cortex.

"There are decisions I would make with a monkey that I would not feel comfortable making with a chimpanzee," said the University of Wisconsin psychologist Chris Coe, who was not involved in the study. "Some of the other cognitive abilities that monkeys would have to show, I don't believe they do. I don't believe they sit and ponder their fate, or reflect on the past, or fret about the future, because they are able to see themselves in a mirror," he said.

"We don't know whether they have a sense of past or future," said Marino, who called Coe's research distinction an ethical non sequitur. "Whether an animal has a sense of the past or future is irrelevant to the issue of whether they can suffer in the present."

Even if Coe accepts human-benefiting research involving contagious diseases or invasive procedures in monkeys that he wouldn't in chimps, however, he said the findings underscore the importance of improving research animal conditions. The macaques' unexpected self-awareness certainly influences the equations by which society must continually balance the harms and benefits of research.

"A study such as this one, that pushes our own awareness of what monkeys can and can't do, challenges us," Coe said. "I'm not going to argue that having animals live in small cages is so wonderful. One has to reflect on that."

A more accurate understanding of animal awareness may ultimately require better tools. Many researchers are skeptical of the mirror test, which Marino said "is shaped more by the cognitive limitations of human researchers than anything else."

Wrote Marino in an e-mail, "Other animals may be more deeply contemplative than humans—we just don't know. That's really the bottom line. Any scientist who tells you they know that other animals don't think as richly or as complexly as humans—is, well, not being scientific."

Author's note: Since publication of Populin's study, more scientists have come to acknowledge that mirror self-recognition is overemphasized as a measure of self-awareness. It's certainly possible to be self-aware without recognizing oneself in a mirror. That said, it remains a measure of what might be an especially pronounced sense of self, not unlike our own—and further research has supported the possibility that monkeys indeed recognize themselves this way.

THE NEW ANTHROPOMORPHISM

My suburban neighborhood is one of those fortunate places inhabited by plenty of animals as well as people. Crows cavort outside my window; down the street is a pond where frogs lurk in the reeds and swallows patrol the surface, and my running path takes me past browsing deer and along a creek where turtles bask and sunfish nest.

I keep an eye out for them all. Animal-watching is like people-watching at a streetside café, engrossing me in imagined stories. To watch a chickadee is to live for a moment in the trees. And when I encounter my nonhuman neighbors, I often wonder: What's on their minds?

It's a scientifically exciting moment to ask that question. For most of the past century, the official answer leaned toward: *nothing much, really.* With a few notable exceptions, scientists defined animals as instinct-driven and incapable of thought, or else governed by simple stimulus-response conditioning. Human intelligence was treated as singular, differing from other animals not in degree, as Charles Darwin wrote in *The Descent of Man*, but in kind. To assert otherwise was to invite the invalidating taint of anthropomorphism: imputing human characteristics to objects that don't have them, not unlike a child playing with stuffed animals. It was unscientific.

How times have changed: what once was considered anthropomorphic thinking is now mainstream science. That's not to say researchers have come to see other animals as simply furred or feathered versions of ourselves. But they are increasingly attentive to the shared biology of human and animal consciousness. A consensus is emerging that to study animals is to appreciate not only their differences from us but also their deep similarities. As the primatologist Frans de Waal writes in *Are We Smart Enough to Know How Smart Animals Are?*, "anthropomorphism is not always as problematic as people think."

Are We Smart Enough is the latest in a profusion of books by scientists and popular-science writers: see also Carl Safina's *Beyond Words: What Animals Think and Feel*, Nathan H. Lents's *Not So Different: Finding Human Nature in Animals*, Jonathan Balcombe's *What a Fish Knows: The Inner Lives of Our Underwater Cousins*, and Jennifer Ackerman's *The Genius of Birds*, all published in the last year. New research describes qualities among nonhuman animals that were once considered exclusive to us: empathy, mental time-travel, language, self-awareness, and altruism. Journals overflow with studies of animal minds, frequently described in language also used to describe the human, and feats of animal intelligence seem to go viral weekly: octopi escaping their tanks and crows gathering to mourn their dead, problem-solving fish and grieving monkeys and sociable snakes.

Not everyone is so excited about it. To some scientists, these celebrations of animal smarts are often held before alternative explanations can be ruled out—"killjoy explanations" that are "less likely to make headlines," as Sara Shettleworth, a University of Toronto psychologist and zoologist, wrote in a 2010 *Trends in Cognitive Sciences* essay. Clive Wynne, who studies animal cognition at the University of Arizona, says that "anthropomorphic descriptions of animal behavior shed more smoke than light." And it's worth remembering that scientific inhospitability to animal intelligence emerged partly in response to the carelessness of many early anthropomorphizers.

Darwin's protégé George Romanes, for example, in the 1880s leapt from secondhand just-so animal stories, such as that of a monkey who supposedly shamed a hunter by proffering his bloodied paw, to conclusions about their mental abilities. Perhaps not surprisingly,

Romanes' own acolyte, Lloyd Morgan, took a sharp turn toward the empirical; his caution—that animal behavior should never be seen as evidence of "higher psychical faculty" if it could be explained by one "lower on the psychological scale"—would become enshrined as Morgan's Canon, shaping a coming century of research.

Although adherence to Morgan's Canon arguably became dogmatic, it's also true that scientists were short on better data and theories. As Shettleworth herself noted, anthropomorphism is now accompanied by scrupulous experiments and sophisticated investigations. These describe not only behavior but also neurobiology and draw on recent research into human psychology and cognition, making it possible to compare the inner workings of animal minds with our own.

The most exciting areas now involve moving from obviously intelligent species, like chimpanzees and elephants, to less-understood animals, and figuring out how complex thoughts arise from the interplay of simpler cognitive processes—how to gauge not simply intelligence but also feeling and inner experience, and what this blooming, buzzing profusion of minds means to *Homo sapiens*.

"The real job has only now begun," says de Waal. "We are now ready to get into the details."

Now sixty-seven years old and holding an endowed professorship at Emory University, de Waal began his career in the late 1970s, not long after the zoologist Donald Griffin's seminal *The Question of Animal Awareness* made waves by suggesting that animals could think and reason. It was a radical idea—Griffin's *New York Times* obituary in 2003 would say he "broke a scientific taboo"—and the research for which de Waal is now best-known, on empathy, altruism, and social intelligence in chimpanzees, was initially received with skepticism. De Waal's findings that chimps reconciled after fights and exchanged grooming for food, and that they did so consciously, was "taken to imply impossibly high cognition," he recalls.

Even studying those behaviors was at first difficult. "I learned to go with mainstream topics to get funding, and to test my more speculative ideas simply in the context of such funded work," says de Waal. "I have never received a penny, for example, for research on reconciliation and empathy, even though this is what I am most known for." Though the quality of his science would eventually prevail, it required both inner conviction and a thick skin.

As difficult as it once was for de Waal and other primatologists, though, their task was comparatively easy: Chimpanzees are humanity's closest living relative, and other primates are still part of the evolutionary family. To recognize their intelligence is, in a sense, to recognize our own. In recent years, chimps have been phased out of most medical research, and the US National Institutes of Health is reviewing how monkeys are used. But science and society have been slower to acknowledge our similarities with more distantly related species.

Take rats: They are widely underestimated, de Waal says, and exempt from Animal Welfare Act protections, even though findings made in monkeys are frequently then demonstrated in rats. In fact, in a series of experiments conducted over the last several years, rats have been found to possess empathy, a capacity sometimes said to define humanity.

That research was first described in 2011 in the journal *Science* by Peggy Mason and Inbal Ben-Ami Bartal, neurobiologists at the University of Chicago and the University of California at Berkeley, respectively, and built on earlier studies of how mice experience "emotional contagion," or a sort of ripple effect in which individuals become distressed when nearby individuals become distressed. So acutely do rats feel for trapped cagemates, found Mason and Bartal, that they preferred helping them to eating, even choosing helpfulness before chocolate—no small thing to a rat. Further research described the importance of emotion to this response. When the scientists gave rats a drug that blocked their ability to feel stress, they stopped helping.

These findings were notable not just for putting a kinder, gentler face on an oft-loathed rodent, but also for underscoring the complexity of empathy. The term encompasses different forms, from the direct response of those rats to humans who are moved to tears imagining the predicament of someone they know only from a story.

Yet much of our everyday empathy does involve caring responses to people in front of us. "Many of the characteristics we value most in ourselves are aspects of our nature that originated in other animals," says Susan Lingle, a behavioral ecologist at the University of Winnipeg, "and are traits that we still share." Being anthropomorphic can mean recognizing the animal in us.

Yet do rats truly experience empathy or do they simply *behave* in a similar way while possessing a very different internal reality? The question poses what's known as the other-minds problem: The only feelings we can observe directly, and thus be sure about, are our own. Without the ability to ask animals to explain themselves, scientists can only do their best to plot a trajectory from multiple lines of behavioral, neurological, and evolutionary evidence.

Sometimes they disagree on where the lines point. This applies to our understanding of all animals, not just rodents, but an especially instructive example again comes from rats and whether they can imagine themselves in the past and future. Technically known as mental time-travel, this is a bedrock part of human experience; in 1997 two psychologists, Michael Corballis and Thomas Suddendorf, declared that it constituted "a discontinuity between ourselves and other animals," whose mental lives were constrained by the boundaries of the present moment.

In coming years, evidence mounted for mental time-travel in other species. Researchers reported conscious planning by chimpanzees selecting tools they'd later need, or western scrub jays caching food for the following day. Corballis and Suddendorf challenged those findings, saying they could be explained by a rote association of action and reward without recourse to high-powered mental projections. Yet over time, Corballis, a professor at the University of Auckland, started to have misgivings.

"I thought we were getting kind of picky in not accepting that evidence and instead trying to pull those experiments apart," he recalls. Meanwhile, studies of rats' brains showed the same neurological networks that in humans underlie mental time-travel and spatial memories. When rats who'd previously earned a treat by navigating a maze later displayed telltale activity patterns in those networks, as if they were actively retracing their steps and plotting a future path, Corballis was swayed.

"I am among those who have claimed that [mental time-travel] is a uniquely human capacity," he wrote in 2013 in the journal *Trends in Cognitive Sciences*, "but I now question whether this is so." Suddendorf, a professor at the University of Queensland, dug in his heels: Whatever the rats experienced, he rejoined in his own *Trends in Cognitive Sciences* letter, it simply couldn't compare with the richness of ours. Invoking the way we "mentally populate spatial scenes with actors and actions, embed scenarios into larger narratives, and reflect on the likelihood or desirability of different options," Suddendorf likened the difference between our experience and a rat's to that between H2O as a solid or a gas. The elements might be the same, but they took radically different forms.

It's presently impossible to adjudicate who is right. Scientists must make judgment calls on those trajectories of evidence, and to Corballis they point unavoidably to mental time travel. Furthermore, he points out, the ability to remember the past and imagine a future has clear evolutionary benefits. He thinks continuing skepticism reflects not objective rigorousness but a deep-seated desire to prove humans superior.

Corballis does grant that our mental time-travel is probably far richer than that of rats. We can expand our mental scenarios with all manner of fantasy and hypotheticals. The rats who live next to a Mexican restaurant's garbage bin probably don't imagine themselves inside with all their friends and extended family, throwing a *Ratatouille*-style party. Still, they might at least think of the delicious treats awaiting them after closing time. It's a difference of degree, not kind.

I think of that while watching crows, a raucous gang of whom rule the airspace between my building and the next block, and who like to store food in hollows at the tops of telephone poles. Maybe, as Suddendorf would have it, they're acting on instinct and basic learning: a combination of behavioral routines reinforced when crows later rediscover the food they've hidden with no conscious thought of the future. Or, per Corballis, maybe the crows actively plan for a time when they'll be hungry.

It's certainly possible. Like scrub jays, crows are corvids, a family of birds celebrated in books like Bernd Heinrich's classic *The Mind of the Raven* and John Marzluff's 2013 encomium, *Gifts of the Crow: How Perception, Emotion, and Thought Allow Smart Birds to Behave Like Humans*. They're avian valedictorians, frequently compared to great apes; at this point, feats of corvid intelligence are old hat. What's more intriguing is the study of thought in other, less intellectually celebrated birds, like the black-capped chickadees who live in a patch of forest just down the street.

They too cache food—hundreds of seeds a day, tens of thousands in fall and winter—the locations of which they remember with marvelous accuracy and rely on to survive. This would at least suggest the possibility of foresight, and indeed some preliminary research supports the notion, though such a capacity might seem beyond their lentil-sized brains. Still, it's careless to assume chickadees are simple. Indeed, research into a closely related bird, the great tit, suggests they possess something else that's quite impressive: the fundaments of language.

One of these is "referential communication," or the ability to make sounds that don't just reflect some internal state, like a gasp of surprise or cry of distress, but refer to something specific and external: a crow, for example, or a rat snake, as described in research conducted over the last several years by Toshitaka Suzuki, an ethologist at Japan's Rikkyo University. Then, in a study published this year in *Nature Communications,* Suzuki and Michael Griesser of the University of Zurich investigated how the meaning of their calls changes depending on the order of sounds. Play a recording of a call sequence that means "scan for danger and come here," and the birds look to the horizon and flock to the loudspeaker, but reversing the chirps occasions no response. Put another way, the birds used syntax.

In human languages, syntax makes it possible to generate ever-more-varied meanings from a limited range of sounds. Scientists have long believed animals lacked both syntax and vocabulary, says Griesser. "The problem was that we didn't listen carefully enough."

So far only one phrase has been deciphered. Many more remain to be translated. If we could understand them, thinks Griesser, we'd mostly overhear conversation of immediate relevance to the birds: "I'm over here." "There might be food." "My territory." "Look out, a hawk!" There are limits, though, to what playback experiments can reveal. It's much easier to decipher "hawk" and "come here," which elicit obvious and easily quantifiable responses, than "How are you feeling?" or "I'm happy to see you."

That sort of chatter could be within the birds' cognitive power, and certainly should be relevant to their highly social life, but would be hard to test experimentally. Which is of a piece with a historical tilt in animal-intelligence studies toward more easily empiricized cognitions: even de Waal shies away from matters of emotion and subjective experience. "Even with humans," he says, "it's a tricky topic," and in animals, "I'm not sure we'll ever get there." To be clear, de Waal's not asserting that rich animal emotions and feelings don't exist. He's just not optimistic about measuring them rigorously.

Some scientists address that other-minds problem by trying to be precise with their terminology. Declaring himself uncomfortable with anthropomorphic language, Hans Hofmann, a neurobiologist at the University of Texas at Austin, prefers to talk of stimuli varying in salience and valence depending on context. He doesn't disbelieve in animal emotion, noting that relevant neural substrates were present in the last common ancestor of all living vertebrates, and that "aspects of affective processing are present in a broad range of species," but he doesn't like the word. It's too vague.

Other researchers are suspicious of the very idea. Joseph LeDoux, a neuroscientist at New York University, sees "a wave of unconstrained mentalism" in language about animal "fear" and "pleasure" and "hunger." Animals might have subjective experiences, LeDoux says, but people shouldn't hasten to understand them in terms of our own. Human cognitive processes may very well produce inner experiences profoundly different from an animal's. The University of Arizona's Clive Wynne thinks our own experiences are shaped in fundamentally unique ways by human-specific cognition, especially—chickadee syntax not withstanding—our rich language.

That's possible, but it's also a hypothesis. Scientists don't know how language shapes human subjectivity; the foundations may well have been laid while our language still resembled that of chickadees, or for that matter Campbell's monkeys or prairie dogs, two other species in whom syntax has been found. Perhaps language *does* shape subjectivity, but in such a way that, as with mental time-travel, the basic components produce something recognizably similar if less sophisticated. And Griesser, who hopes to quantify the strength of chickadee affections by measuring how much time they spend watching out for their mates, challenges the very premise that language is of prime importance. "I think that we humans don't think about emotions with words," he says, and instead of articulating them to ourselves linguistically rely on feelings and wordless thoughts.

The ethologist and author Marc Bekoff thinks primary emotions—including fear, joy, happiness, jealousy, anger, love, pleasure, sadness, and grief—are probably widespread and that hewing too closely to abstract description could become a form of what he calls "anthropo-denial." Take the neuroscientist Jaak Panksepp's experiments in the late 1990s with rats subjected to "playful, experimenter-administered, manual, somatosensory stimulation": If they emit vocalizations associated with positive emotional states, why not say—as Panksepp did—that they laugh when tickled?

It's hard to imagine rats laughing or chickadees conversing without also imagining they're self-aware. Trying to conceive of experience absent a sense of self is practically a Zen koan: What is you if there is no *you*? In his book *Not So Different*, Nathan H. Lents, a molecular biologist at John Jay College of Criminal Justice, defines animal self-awareness as "a sense of themselves as distinct from other individuals and within the context of their environment and circumstances." It does not sound like a high bar to meet, but the idea remains contested.

Only a few species, including dolphins, elephants, orangutans, magpies, and most recently manta rays, are able to recognize themselves in a mirror, which since the 1970s has been considered a foundational test of self-awareness. Human children usually pass it by the time they're three years old. Yet some researchers say reliance on the mirror test is misguided. Lori Marino, formerly a neurobiologist at Emory University who in 2001 was part of the research team that demonstrated dolphin self-awareness, thinks the emphasis placed on the test has made it a tool of exclusion: animals who don't demonstrate our own particular form of self-awareness are denied the capacity altogether.

"We still use ourselves as a measuring stick," says Marino. "If an animal is sufficiently like us, only then are they worthy of being valued." De Waaal shares some of her misgivings, arguing in *Are We Smart Enough* that *every* animal "needs to set its body apart from its surroundings and to have a sense of agency." Mirrors measure only one form of self-awareness, he says; many animals spend their entire lives without ever encountering a reflection and may rely on other cues, such as smell.

"This is certainly the case in fishes," wrote Culum Brown, a behavioral ecologist at Macquarie University, in a 2014 review of fish intelligence published in the journal *Animal Cognition*. "Chemical cues play a very important role in aquatic ecosystems. There is compelling evidence that fish are capable of self-recognition using chemical cues." Even fish recognize themselves and other individuals, then—creatures whose expressionless faces and alien habits so predispose people to thinking them unfeeling that even "vegetarians" eat them.

Fish are an especially interesting example of animal cognition studies expanding not only beyond species of evident intelligence but also beyond entire evolutionary groups and into parts of the animal kingdom where many scientists remain reluctant to posit similarities with humans. Or, to put it casually: Sure, we have some things in common with mammals and birds, but *fish?*

That tension was on full display in a recent debate held in the journal *Animal Sentience* on whether fish feel pain. The argument that they don't, championed by Brian Key, a neuroscientist at the University of Queensland, rests on neurological differences: fish can have aversive experiences, but their brains are just too different to produce the emotional resonance crucial to what we understand as pain. Brown and others responded that, despite the anatomical dissimilarities, fish brains and chemistries could still generate something similar, and that behavioral studies certainly showed them acting like they felt pain. Key rejoined that fish brains are still poorly understood and the behaviors overinterpreted.

In some ways the debate felt like a litmus test for how different researchers weigh evidence. George Striedter, a neurobiologist at the University of California, Irvine, took a middle ground: "The debate cannot be settled yet," he wrote, and he noted that fish have demonstrated many other cognitive feats. These include what seems to be fear; the ability to count, as described in angel fish who differentiate between schools with different numbers of individuals; physiological stress responses considered markers of consciousness in other species; long-term memories; and even culture, described experimentally in three-spined sticklebacks who learn feeding strategies from their compatriots.

Pain aside, there's clearly some extraordinary cognition going on underwater—and while it can be misleading to talk generally of fish, of which there are roughly thirty thousand species, Brown thinks that that findings from model organisms, like those made in zebrafish or sticklebacks, can be extrapolated broadly. Gordon Burghardt, a University of Tennessee, Knoxville, ethologist who in the 1980s coined the term "critical anthropomorphism" to describe how scientists could formulate hypotheses about other species by trying to imagine life from their perspective, has long lamented that turtles and other reptiles are, like fish, underestimated by virtue of their seemingly expressionless faces and unfamiliar habits; many basic aspects of reptile life history remain unknown, and research in turtles has been skewed by a tendency to notice mostly what they do out of water, ignoring the social interactions that occur beneath its surface.

In the 1990s, Burghardt described turtles playing—or, in his careful language, engaging in "incompletely functional behavior differing from more serious versions structurally, contextually, or ontogenetically, and initiated voluntarily when the animal is in a relaxed or low stress setting." (Komodo dragons also seem to play; sped up on video, Burghardt notes, they look like dogs.) More recently he and the then-doctoral student Karen Davis also found evidence of long-term memory in American red-bellied turtles, as well as an ability to learn by watching others. What might *their* emotional lives be like? "Although the nature of subjective experience is only partially accessible to objective science," Burghardt wrote this year in *Animal Sentience*, "we must keep trying to understand it." Certainly neurological networks fundamental to social interaction and emotional reward are present in them, as they are across all vertebrates.

For Burghardt, who remembers "when you couldn't even talk about animals being hungry because it was too anthropomorphic," the present moment is a tremendously exciting one. "In the period right after Darwin, scientists were looking for commonalities between humans and other animals, but they didn't have good tools or experimental designs," he says. "Now people are doing experiments on phenomena that 20 or 30 years ago nobody even looked at. They're finding abilities that we never thought animals would have."

In recent years scientists have even found that insects possess evolutionarily ancient brain structures responsible for creating mental maps of one's own place in space. Some researchers consider these structures foundational to human awareness; if they are, then insects, too, would appear to be conscious. Whatever it feels like to be a bee, it feels like *something*.

What that something is, how instinct and awareness interact, how different forms of memory shape experience, how evolution's convergences and divergences have shaped the development of cognition across time and circumstance—these are frontier questions now being asked. Science has come a long way from a reflexive adherence to Lloyd Morgan's wariness of "higher psychical faculty" or the famed behaviorist B. F. Skinner's insistence that other animals are "conscious in the sense of being under stimulus control" and experience pain with no more conscious resonance than "they see a light or hear a sound."

Other questions involve capacities like morality: Might its biological building blocks be widespread in the animal kingdom? Or what about motivation: after all, a human whose every physical need is provided for, but who doesn't actually do anything except sit in a room, won't be very happy. Beyond seeking pleasure, avoiding pain, and procreating, what might an animal find fulfilling?

"I don't think I can understand that unless I try, with a whole lot of humility, to imagine what it would be like to be that animal," says Becca Franks, a cognitive psychologist at the University of British Columbia. "Then you take those insights to create an experimental, data-driven paradigm. That's how science proceeds."

Franks sometimes has misgivings about the emphasis placed on experimentation with animal brains. Even as scientists learn more about the depth of their thoughts and feelings, the animals themselves are often treated as a means to our own ends. It's a concern shared by Lori Marino, whose research led to her present role as animal advocate and activist, marshalling the science to argue for better ethics. There's a limit—or at least *should* be a limit—to what we can study. "There's a lot of stuff I'd like to know," says Marino, "but I'm not going to kill a dolphin or stick electrodes in their brains or do invasive work just because I want to satisfy my curiosity."

Yet a great deal can still be learned in ethical ways; as with research on humans, ethical constraints can spur greater ingenuity. And as this research proceeds, Frans de Waal has called for a "moratorium" on claims of human uniqueness. "We are still facing the mindset that animal cognition can be only a poor substitute of what we humans have," he writes in *Are We Smart Enough*. Only by setting aside that preconception for a few decades may we "then return to our species's particular case" with a deeper basis for understanding what truly distinguishes us.

As for what to make of the animals I encounter while running along the creek, I think of research by Susan Lingle at the University of Winnipeg. She describes how white-tailed deer mothers respond to distress calls from infants belonging to other mammal species, including marmots and cats and *Homo sapiens*. It's an automatic behavior, says Lingle, but likely accompanied by conscious awareness and concern. Those moms care about the distressed.

It calls to mind another line of research, on what's called "mental state attribution." If those deer mothers can respond to a human baby's cry, I wonder, might they be imagining how that baby feels?

As described last year in the journal *Animal Behaviour* by two University of Vienna behavioral biologists, Esmeralda Urquiza-Haas and Kort Kotrschal, mental state attribution refers to the tendency of animals to project thoughts and feelings onto other animals. The cognitive underpinnings of this, they say, are found across the animal kingdom. It seems to have been an evolutionarily favored way of making sense of the living world.

We're biologically predisposed to anthropomorphize, then, as are other creatures in their own species-specific ways. Turtles may turtle-pomorphize, frogs frog-pomorphize, and so on. Even as I'm watching a mother deer, making sense of her mind in terms of my own, she's probably doing the same to me.

HONEYBEES MIGHT HAVE EMOTIONS

Honeybees have become the first invertebrates to exhibit pessimism, a benchmark cognitive trait supposedly limited to "higher" animals.

If these honeybee blues are interpreted as they would be in dogs or horses or humans, then insects might have feelings.

Honeybee response "has more in common with that of vertebrates than previously thought," wrote the Newcastle University researchers Melissa Bateson and Jeri Wright in their bee study, published in *Current Biology*. The findings "suggest that honeybees could be regarded as exhibiting emotions."

Bateson and Wright tested their bees with a type of experiment designed to show whether animals are, like humans, capable of experiencing cognitive states in which ambiguous information is interpreted in negative fashion.

Of course, unlike unhappy people, animals can't say that the glass is half-empty. Researchers must first train them to associate one stimulus—a sound, a shape, or for honeybees, a smell—with a positive reward, and a second with a punishment.

Then, by prompting the animals with a third, in-between stimulus, it's possible to assess their outlook. Like a depressed person seeing hostility in a neutral gaze, pessimistic animals tend to treat that uncertain stimulus like a punishment.

Such tests might seem simplistic compared with the richness of human emotion, but they're the most objective available tool for comparing cognition across species. And pessimism is no mean feat: It's a form of cognitive bias, considered in humans to be an aspect of emotion. You can't be pessimistic if you don't have an inner life.

Earlier research has found rats and dogs capable of pessimism. Bateson has also documented pessimism in starlings. But though honeybees have passed tests of pattern recognition and spatial modeling, the idea of feelings occurring in their sesame-seed-sized brains is generally considered unlikely, if not downright laughable.

"Invertebrates like bees aren't typically thought of as having human-like emotions," said Bateson, yet honeybees and vertebrates share many neurological traits. "Way, way back, we share a common ancestor. The basic physiology of the brain has been retained over evolutionary time. There are basic similarities."

Until now, though, they hadn't been tested. Bateson and Wright trained their honeybees to associate one scent with a sugary reward and another scent with bitterness; then they shook half their beehives, mimicking a predator attack. Afterward, shaken bees still responded to the sugary scent but were more reluctant than their unshaken brethren to investigate the in-between smell.

Further analysis of the shaken bees' brains found altered levels of dopamine, serotonin, and octopamine, three neurotransmitters implicated in depression. In short, the bees acted like they felt pessimistic, and their brains looked like it too.

"The methodology is sound," said Lori Marino, an Emory University evolutionary neurobiologist who was not involved in the study. "I don't think it's a stretch to say that they are tapping into bee emotions. After all, every animal has to have emotions in order to learn and to make decisions. And we already know from many other studies that bees are really cognitively sophisticated."

But Bateson said the results could be interpreted another way. "Either bees have feelings, or cognitive bias isn't as tightly correlated with feelings as we thought," she said. "Maybe cognitive bias is not a good measure of emotion."

In future studies, Bateson hopes to elicit from honeybees other forms of apparent emotion, such as happiness. She also wonders about the mental effects of chemicals and disease.

"It would be interesting to know if pesticides were altering their cognition, creating states similar to depression," she said.

Author's note: While preparing this book, researchers led by Clint Perry at the Queen Mary University of London published findings similar to Bateson's and Wright's, only using visual rather than scent cues, and emphasizing not pessimism but optimism. It's not unreasonable to walk past a bank of flowers abuzz with bees and think of them as—well—cheerful.

III

INTERSECTIONS

Most scientists I know have a deep, abiding faith in the power of knowledge. If only people had more information, they think—if only there was better data and methods, a clearer communication of fact—then progress would follow.

It's a noble belief. It's what makes them who they are: people who ask questions for a living, and whose empirical rigorousness and ingenuity in answering those questions makes science so wonderful. It's also naive. Few problems today stem from lack of knowledge. Rather, they're rooted in a failure, for whatever reason, to apply it. Someone needs to actually do something.

My own habits of mind are not much help in this regard. I've an unfortunate tendency, when I learn about the broken ecology of dammed rivers or the suffering of captive chimpanzees, to become frustrated and judgmental and hectoring. Thankfully not everyone is like this; they work to pull down those dams, to change the legal status of chimps, to make the world a better place by applying scientific insight to the way we live. And it's my privilege to write about them.

A DAY IN THE LIFE OF NYC'S HOSPITAL FOR WILD BIRDS

Even in a city famed for its oddities, New York's Wild Bird Fund is an unusual place. To wit: one morning in April, while sitting in their Upper West Side waiting room, it dawned on me that I wasn't alone. Perched on a chair, motionless in front of a life-sized photograph of a turkey vulture, was a large black-and-white guinea fowl.

The turkey vulture had been nicknamed Stanley. The guinea fowl, who'd been let out to stretch her wings, didn't yet have a name. Both were among the roughly ten thousand feathered patients—snowy egrets and starlings, peregrine falcons and pigeons—delivered to the Wild Bird Fund since its founding in 2001.

"Any animal that's picked up is in really bad shape," said Rita McMahon, the Wild Bird Fund's founder. "Half the animals brought in for rehab die, or are euthanized. But the other half go free. And they probably would have died if they hadn't come here."

McMahon founded the Wild Bird Fund after rescuing an injured Canada goose alongside Interstate 684 and learning there was nowhere in the city to take him. New York is full of veterinarians, of course, but none wanted to deal with wild animals. At first she ran the operation from her apartment. "I didn't have any idea," she said. "I just started doing it."

It was a far cry from the current facilities, through which she gave me a tour. On the first floor is an intake room for initial observations, a quarantine chamber for contagious birds, a microscope and an X-ray machine, and a wading pool for waterfowl. Downstairs you'll find the operating room and cages for the patients.

Recent patients included a Virginia rail and a woodcock. For now, aside from a yellow-bellied sapsucker and dark-eyed junco in intensive care, both victims of window collisions, it was mostly pigeons. Medical care for lowly pigeons might seem a bit much, but it was hard to begrudge after hearing their tales.

McMahon pointed me to a climate-controlled cubicle containing two nestlings, the only survivors of a brood of thirteen that had been packed into a FedEx box. "A guy thought it was a good joke to send them to his girlfriend," she explained. "They were there in the box for two days with no water, no food, and freezing. They all looked dead. When we heated them up and gave them fluids, these two made it. The girlfriend broke up with him."

Upstairs it was time for the waterfowl to take their daily swim in the wave pool. Without it, their feathers lose their natural waterproofing, and their feet dry out and become infected. Sometimes a loon comes through; they're allowed to dive for goldfish. Mute swans are frequent guests. The males are renowned for their antagonism toward men, and accept care only from the hospital's women helpers.

On this day the waterfowl included two Canada geese, three herring gulls, and a mallard. Under the care of the volunteer Esther Koslow they took their laps, then gathered in the waiting room, preening quietly.

Downstairs it was rather less peaceful. After feeding, while their cages were cleaned, several pigeons stayed out for exercise. Assisting were several students from the LaGuardia Veterinary Technology Program, none of them especially good at catching pigeons. Mild pandemonium ensued.

"No badminton!" McMahon shouted, referring to one of the trainees' pigeon-catching attempts. "Talk to the bird. 'Hello, you beautiful thing,'" she cooed, plucking one from midair. With Derek Jeter-like aplomb, she snagged another with her other hand, then returned both to their cages.

McMahon wasn't annoyed. After all, it's a small army of volunteers who make it possible to accept, free of charge, every bird that comes through the door. More than two hundred people help, most having learned of Wild Bird Fund after finding an injured bird themselves.

The volunteers span all walks of city society. Homeless people distribute flyers. The antique dealer Stephanie Rinza, part-owner of the Vanderbilt mansion, recently let the Wild Bird Fund use it for a fund-raiser. At the party, said Koslow, a one-eyed male cardinal fell head-over-tailfeathers for Sigourney Weaver. She wore a dull orange dress and may have resembled a large female cardinal.

With the pigeons cooped, several larger fowl roamed the floor: the guinea hen, a large red rooster, and a chukar, a small gamebird that is the national bird of Pakistan. It was found in Queens. They'd all most likely escaped from live poultry markets. Different as they looked from one another—the chukar barely reached the rooster's knees—they're members of the galliform order, and instinctively clustered together.

"Chickens, chukars, guinea hens—these are not part of our mission, but we take them all in," McMahon said. "What else are you going to do?" Wild birds are returned to the wild, but these would probably be sent upstate to a sanctuary run by Zezé, the eponymous founder of Zezé Flowers, one of the city's best-known florists.

When wild birds need extra time to recuperate, said McMahon, they go to the Raptor Trust in New Jersey. Raptors are regularly brought in, including a female red-tailed hawk that had stepped in tar and calmly allowed a volunteer to scrub it from her talons. "As soon as we put her in a cage, she was ferocious," recalled McMahon. "But while we cleaned her, she didn't even stir. It was like the story of the lion with a thorn in his paw. She knew what we were doing."

As the day progressed, a worker from the city's Department of Parks & Recreation arrived with an escaped lovebird. A young man dropped off a pigeon found immobile on his apartment building's roof. Eugene Oda, another volunteer, looked him over. The prognosis wasn't good. Much of their work, said McMahon, is simply providing birds with comfort in their final hours: food, water, sedatives. It's avian hospice care.

Oda went to the operating room to resplint a pigeon's leg. The splint was made from a paper clip and wrapped in medical tape. Later in the afternoon a woman came in with a starling she found on 73rd Street, on her way to a Passover dinner. The starling flapped feebly. A bad sign, Oda said, but he still had a chance to live.

"It was just dying on the side of the road," the woman said tearfully. She stroked the starling's feathers and whispered soothingly. "I don't know why I'm so upset, but I am," she said. "I'm not even a bird person."

NEW YORKERS IN UPROAR OVER PLANNED MASS KILLING OF SWANS

In the menagerie of human mythology, mute swans occupy a special place. Across millennia they've symbolized transformation and devotion, light and beauty. Now a plan to eradicate the birds from New York has made them symbols of something else: a bitter and very modern environmental controversy.

The debate swirls around a host of prickly questions. Are mute swans rapacious destroyers of wetlands, or unfairly demonized because they're not native to this area? Are some species in a given place more valuable than others, and why? Which deserves more protection, the animals that inhabit our landscapes, or the ones that might thrive in their absence? The answers depend on whom you ask.

"Mute swans always attract controversy, and tend to polarize people. And, as with most things, the truth probably lies somewhere in the middle," said the waterfowl researcher Chris Elphick of the University of Connecticut. "The real issue is that there are no simple answers."

Mute swans are not native to North America. New York's population descended from escapees imported for ornamental gardens in the late 1800s. Weighing up to 40 pounds apiece, they can eat 10 pounds of aquatic vegetation daily. In their absence, that food might be eaten by native wildlife. Mute swans are also aggressive during nesting season and have been blamed for attacking ducks and pushing out other waterfowl.

Late in January, New York's Department of Environmental Conservation issued a draft of a plan to reduce New York's wild mute swan population to zero by 2025. Nests and eggs would be destroyed; a few adults might be sterilized or permitted to live on in captivity, but the rest would be killed.

"They are large, destructive feeding birds, and as much as they are beautiful, they can wreak havoc on the underwater habitat that a lot of other fish and wildlife depend on," said Bryan Swift, a DEC waterfowl specialist and lead author of the plan. "We have an obligation to sustain native species. The question then is, 'At what level?' But in the case of introduced species, I don't think we have that same obligation."

That's one way of looking at it. "We have so little opportunity to experience wildlife in New York City," said David Karopkin, director of the animal advocacy group GooseWatch NYC, "and now they're targeting the most beautiful animals that we do have. The fact of the matter is, they're part of our community."

Were mute swans not so beautiful, the plan might not have caused much outcry. But unlike other animals tagged as invasive and pestilential—like Burmese pythons, feral hogs, and snakehead fish—mute swans are widely beloved. For people who live near wetlands around Long Island and the Hudson Valley, where most of New York's mute swans live, they're also a part of everyday life.

To their defenders, the fact that mute swans are nonnative carries little weight. As one Queens resident told the *New York Times*, "If they were born here, they should be considered native by now." And some think the swans' heritage is being held arbitrarily against them.

"If Mute Swans were native to North America, they would not be viewed negatively by state wildlife agencies," said the ornithologist Don Heintzelman, an author of bird-watching field guides who is working with Friends of Animals, a New York-based animal advocacy group, to oppose the plan. Complaints that that mute swans harm other wildlife "are greatly overblown," he said.

Over the last few years, some scientists have argued that non-native species are unfairly persecuted, their negative effects too frequently assumed rather than conclusively demonstrated. In fact, the evidence for ecological harm by mute swans is somewhat mixed.

In a review of mute swan impacts on wetlands published this month, French ecologists said the swans sometimes attacked other birds—but sometimes they did not. Likewise, their feeding habits sometimes damaged aquatic plant communities, but not always.

"The consequences of mute swan presence may strongly differ from an ecological context to another, so that no simple rule of thumb can be provided," wrote the researchers. They did, however, say that the risk of mute swans reproducing prolifically and outcompeting other species could justify their eradication from North America "as a safety measure."

Swan defenders say the need for a safety measure is far from clear. In Maryland's Chesapeake Bay, where a controversial mute swan eradication program was enacted in 2003 to help restore aquatic vegetation, the US Fish and Wildlife Service previously said that swans had likely done little harm to seagrass beds.

"That's a good example of how the science on this is incomplete," said Brian Shapiro, the New York State director of the Humane Society of the United States (HSUS). Whatever effects New York's 2,200 swans may have, he said, is a drop in the bucket compared with human-generated pollution and habitat destruction.

But scientists argue that the effects would be much more severe if mute swan populations grow as anticipated, and more difficult to control. "The dilemma wildlife managers face is that if they wait until there is no doubt that native species are being adversely affected, it will be much, much harder to do anything about it," said Elphick. And although human impacts on habitat and native wildlife are undoubtedly greater, mute swans could consume a disproportionate amount of resources.

"Introduced species are a major cause of extinction and biodiversity loss, and the concern is that if they're not controlled then we will see the world's biota become much more homogenous," Elphick said. "Mute swans are just one example of many."

Mute swans' lives are not the only ones that matter, said Michael Schummer, a waterfowl ecologist at the State University of New York at Oswego. People care so deeply about mute swans because they're aware of them—but if they paid more attention to birds they impact, they might care about those as well.

Author's note: The fight over the fate of New York's mute swans did not abate. The Department of Environmental Conservation revised its original plan to include more nonlethal control methods; the swans' defenders continued to argue that evidence of their harms was lacking in the first place. Recently I asked Don Riepe, director of the American Littoral Society's Northeast chapter and official Jamaica Bay Guardian, whether some compromise might yet be found. "Addling eggs to keep the population down is a good, middle-ground approach," he said. "They should be controlled in national and state refuges but left alone in places like Brooklyn's Prospect Park, where they cause little harm."

AN EEL SWIMS IN THE BRONX

In the annals of natural history, there is perhaps no fish so singularly unusual, even mysterious, as *Anguilla*, the eels. Unlike every other migratory fish on Earth, they spawn in the open ocean and mature inland, in lakes and streams—an elementary fact, yet it took centuries for scientists to discover. They didn't realize that larval eels, which resemble translucent cherry leaves, were actually eels.

Even with this knowledge, nobody has ever seen eels mate. Instead they've inferred its occurrence from the oceanic distribution of larvae. The youngest American eels, or *Anguilla rostrata,* are found in the Sargasso Sea, part of the northern Atlantic Ocean near Bermuda. From there they drift on the Gulf stream up the eastern coast of North America, eating plankton and becoming more eel-like, though still translucent. At some point they encounter, hundreds of miles from shore, a part-per-trillion chemical trace of fresh water and instinctively follow it back to the source.

"They get their cue and detrain," says George Jackman, who one sunny morning in early April is standing in a small park in the southeast corner of Bronx Park, just north of 180th Street, sewing a length of net into a makeshift eel ladder. Jackman, a former New York City police lieutenant turned doctoral biology student, estimates the distance to the Sargasso Sea at 1,603 miles: 1,600 to the Bronx River's mouth and another 3 up the river to 180th Street, where the eels' irresistible migratory imperative meets a 14-foot-tall dam.

Jackman, whose clipped, crime-drama speech assumes a tender note while discussing "probably the most fascinating creatures I've ever studied," wants to help the eels climb that dam. To make the ladder, he's rolled the net into a long ribbon, its form held by rough stitches. Jackman does the sewing; his assistant and two volunteers from the Bronx River Alliance, a local environmental group, hold the net so he can work. They snicker at his incongruous domesticity. Compact and muscular, with close-cropped blonde hair, wraparound shades, and block-like forearms, Jackman looks like he might be off-duty rather than retired. For thread he uses a rope. For a needle he raided his wife's kitchenware. "I pulled the tip off a spatula from Williams-Sonoma," he says. "It's so useful."

The team carries the net past baby-strolling moms and morning park-sitters and out onto the dam. Upstream the river recedes into forest. Downstream it passes into the red brick jungle of the Bronx, population 1.4 million. At the top of the dam, Damian Griffin, the education director of the Bronx River Alliance, picks up a branch. "Signs of the beaver," he says, pointing to the telltale gnawed tip. José, as the borough's sole, beloved *Castor canadensis* is known, was spotted in 2007, after his species' two-hundred-year absence from the watershed.

José soon became a symbol of nature's restoration and resilience within the river, which over the centuries had experienced a full range of urbanization's ecological insults. Its tributary brooks were diverted into sewers. Its banks, except for parkland surrounding the Bronx Zoo and New York Botanical Gardens, were denuded. Well into the twentieth century it was a receptacle for industrial waste, and for raw sewage well into the this century. Technical documents refer to the "Bronx River Sewershed." Yet there is the beaver stick, and a blue heron flying low over the water, and a red-tailed hawk overhead, chased by a blackbird.

The story isn't so happy for migratory fish, though. Whereas a few eels might wriggle up a muddy path beside the dam—they can breathe through their skin, and survive brief sojourns out of water—for most the dam is a wall, cutting them off from waters that might still nourish them to maturity. To make matters worse, gender is a function of population density; eels arriving at the dam are sexually indeterminate, and if they remain crowded at the bottom will almost exclusively become males, further winnowing their populations. The Bronx River situation is not unique. Dams are ubiquitous, and total American eel populations have plummeted to a few percent of historical levels.

The species hangs by a thread, but a thread is still something. Last year Jackman and the biologist John Waldman, the elder statesman of New York City's marine biology, found baby eels at the dam's base, using an eel trap secured by rebar and washed-up plastic bottles. (Jackman also found a 10mm Glock handgun, discarded in the shallows. He brought it to the local precinct, where it was traced to a shooting weeks earlier.) Hence this year's ladder. "We don't know if it will work, but we've got to try," says Jackman.

His team picks a spot on the dam's far side, just below a figurine of Saint Barbara, one of the patron saints of Santeria—a syncretic form of Catholic, West African, and Caribbean faiths popular in the area—along a concertina-wired fence that borders the Bronx Zoo's Wild Asia exhibit. No animals are visible today, but a Wild Asia monorail car and its bemused operator pass several times. The operation takes a couple hours: The netting is threaded through PVC pipe, which is positioned over the dam's lip with rebar and two-by-fours. The net's end rests in a pool below. The pipe channels a trickle of water into the net. Hopefully the eels will detect this and start to climb. A cheer goes up as the water goes through.

"They'd swim all the way to Lake Ontario if they could find a way," says Jackman as he walks back across the dam. Should he succeed, these eels won't get quite that far: there are still two more dams on the Bronx River. But at least they'll have a few more miles of water, a better chance at growing older and fatter, at eventually swimming back over the dam and down to the Sargasso and starting the cycle of eel life anew. The job done for today, Jackman sits on a bench and removes his waders. "This is God's work, if there is such a thing," he says. "You sew the holes in the fabric of this world."

Author's note: In the end, eels didn't use the net ladder; its wavy shape was perhaps too confusing for them. But the effort was still fruitful. Since their low-end fix didn't work, said Waldman, eel advocates pressed their case for a properly engineered ladder. This was installed and is working well.

ON WALDMAN'S POND

"The water," said John Waldman, "varies between green and shockingly green."

Perched on a thin strip of grass between a road and the water's edge, he stared intently at the surface. On a postcard-fall noon it was the color of fresh spinach, the algae and silt so thick that the sun was swallowed just a few inches into the murk.

"Minnows," he said, pointing to a dappling of translucent silver-pink fish several feet from shore. "It's salty, with a sky-high pH, but it's rich and full of life nonetheless." We stood at the mouth of a short creek between two bodies of water that elsewhere would be called ponds, but in New York City they are called Willow Lake and Meadow Lake, and are faintly miraculous. Like so much of the estuary now entombed beneath the world's eleventh-largest city, it was once a tidal marsh, and still receives the tide. This explained the salinity. The pH—nine, to be exact, same as baking soda—came from coal ash, which residents piled by the creek when Queens was still country. Builders used the ash to line the lakebeds, and it prevents the water from spreading back into the silt.

Waldman, a biology professor at nearby Queens College, hoped the minnows would attract a snakehead. The voracious intruder was found this spring in Willow Lake, likely introduced by an owner exasperated with its boundless appetite, and earlier this fall Waldman saw a juvenile—evidence that they might have spawned.

"It's still not clear whether they are highly successful," he said. As for whether the snakeheads will eat everything in sight, fulfilling the fearsome though overstated reputations—Snakehead!!—spread by toothy tabloid covers three summers ago, when the air-breathing Asian natives were discovered in Maryland, nobody knows. "It's possible they'll overshoot the food source before reaching some sort of equilibrium," said Waldman.

Overshooting the food source, of course, is a technical way of saying, eat most everything in the lakes. Worse yet, a breeding population of snakeheads would threaten any other waters to which misguided bucket-carriers could haul them. The Department of Environmental Conservation has taken control of the situation, but Waldman, who worked for twenty years at the Hudson River Foundation before taking his professorship, still comes by to look. He is a fish junky, a city boy who spent his life exploring, and later chronicling the history of, the waters around New York City—waters that, he and others discovered, are full of life, a life that is surprisingly resilient, clinging to and even thriving in niches shaped by human destruction.

Waldman crossed the street to look for snakeheads on the other side, easily vaulting a concrete divider. At fifty he is still fit and trim, with sharp blue eyes, an angler's grip, and a chin of stubble that matches the white of his hair. He wore a brown jacket—herringbone, appropriately—and loose slacks, somewhat resembling the Victorian gentleman who, in a framed drawing on his office wall, carries with aplomb a man-sized fish on a stick over his back.

"The drive for life is really intense. There's an awful lot of contamination in this world," he said, speaking for a moment of life above the water as well. "They may not produce so many eggs as in a pristine environment, or grow as well, but there's still a living to be made here."

Waldman continued up the creek. The leaves had already turned, and many of the fallen had yet to lose their color; thickets of head-high phragmites lined the water's margins with green stalks and loose flaxen heads that waved back the sun's gold. We passed beneath the interleafing of the Van Wyck Highway and the Long Island Expressway, their supporting concrete columns aged by weather and almost as wide around as some of the trees that once lived in the surrounding valleys.

The water's surface vibrated from cars passing above and roiled with schools of minnows who surfaced in flight from our footsteps. A chemical skein floated with lazy iridescence beside a drainage inflow where we next stopped. Littering the water was a collection of trash straight from central staging: milk crates, tires, shopping carts.

"Shopping carts are one of the major features of urban aquatic environments. Another one is the spare tire," Waldman said. Beside one bridge, he said, it is possible to walk across the water on the trash collected beneath it. "Tumbleweed," he said, pointing to a wind-blown plastic bag.

We saw no snakeheads there, or farther along at the locks that hold the lakes in. Nor did we see any other fish, though the water teems with them—sunfish, white perch, carp, catfish, the earlier-seen minnows, which are technically named killifish, and American eels, which despite their name breed in the Sargasso Sea, not far from the West Indies. These fish have survived, even thrived, amidst some of the worst pollution in the world, but now face eradication from the latest offhand transformation.

The snakehead, of course, just a few generations removed from some brackish backwater in northern China, is an unwitting victim in the whole affair, arbitrarily moved by the fate of globalization and human nature. Singling it out as an alien in an already-altered ecological balance could seem unfair—but shaping the development of nature is a power that, for better or worse, people exercise simply by existing.

"The snakehead is a gritty fish, with great survival instincts. It's hard to dislike it," John said on the walk back to the car. "But it just doesn't belong here."

Author's note: The much-feared snakehead apocalypse has not arrived, and new research suggests it might not. In sections of the Potomac River watershed, near some of the first—and most fearfully reported—colonization sites, population increases have slowed and even stopped; in Willow and Meadow lakes, researcher say the snakehead population has grown only slowly, causing no evident harm to local species.

THE RETURN OF THE RIVER

Back when I was a kid growing up in Bangor, Maine, a small city on the banks of the Penobscot River, my father and his fishing buddies fought an effort to rebuild a dam right outside of town. The structure dated back to the 1870s but had crumbled over the intervening century; a hydroelectric company named—with timeless corporate irony—the Swift River Company wanted to bring it back. I was too young to attend the hearings, but to this day the memory of Dad's anger remains fresh.

My father was a member of the Penobscot Salmon Club, one of three gentlemen's fishing clubs on Maine's largest river, which is home to the last Atlantic salmon population of any consequence in the United States. The clubs were buzzing social institutions in their heyday—somewhere in Dad's old film slides is a photograph of my eventual high school history teacher, riding a motorcycle bare-chested with a fly rod between his teeth—but salmon were the attraction for him and his friends: big and elusive and physically beautiful, even beloved.

That might seem like an odd word to attach to a fish, but the anglers regarded those salmon with reverence. They represented a Platonic ideal of fishness; it was even traditional to send each season's first Penobscot salmon to the president, though that custom would soon end. By the mid-1980s, when the fishermen fought to stop the new Bangor dam, salmon populations were already in sharp decline.

All sorts of factors were to blame, from commercial overfishing to industrial pollution, but dams on the Penobscot and its tributaries were the greatest culprit. Atlantic salmon are migratory, born in freshwater streams, traveling hundreds of miles to grow up in the northwest Atlantic, then fighting river currents to return to their home waters again, where they spawn and die. Like clots in arteries, dams block this circulation. Salmon born in most of the Penobscot's 2,400-square-mile watershed had to pass through a gauntlet of four dams on the river's main stem: one in Veazie, just above Bangor, and three more over the next 30 miles.

The Penobscot is hardly unique. For every 100 miles of river in the northeastern United States, there are an estimated seven dams. The largest were constructed in the twentieth century to harvest electricity, but many date back centuries, used to power mills or collect timber-company logs floated downriver. The biologist and historian John Waldman writes that the Penobscot and other northeastern rivers once "ran silver" with migratory fish—not just salmon but also eels and herring, rainbow smelt and striped bass and sturgeon. Today, writes Waldman, dams have left those rivers "fettered and tired, falling to the sea with the stilted tempo of the subjugated."

As far as biologists can figure, fish populations in northeastern rivers now number between 1 and 10 percent of preindustrial levels. It's estimated, for example, that 20 million "river herring"—as alewives, blueback herring, and American shad are collectively known—used to swim up the Penobscot each spring. This year, in the trap where biologists count fish swimming past Veazie dam, they counted 14,000, less than 1 for every historical 1,000. Even that drop-in-the-bucket number was largely possible because of restocking efforts by Maine's Department of Marine Resources. John Waldman calls these decimated populations "ghost fishes."

My father's fight, then, against the Bangor dam—and another, even more bitter battle shortly thereafter, over the proposed Basin Mills dam—were rearguard actions, keeping a bad situation from becoming even worse. Not long after, Dad grew tired of fishing and fighting, and as I grew older, the Penobscot receded from both our lives. Every now and then we would talk about the latest salmon counts, or he would forward a newspaper article about the river, but there wasn't much to say. The old dams weren't going anywhere. They were like features of the landscape, immovable as the river's stone bed. Early this millennium, Dad heard something about environmentalists who wanted to purchase and remove them, at a cost of roughly $40 million—a sum so vast in the context of pot-luck fund-raisers and Maine's working-class budgets as to seem absurd. He didn't pay it much mind. After he passed away in 2009, I didn't think of it again.

Until last winter, that is, when I learned that the removal Dad had considered a fantasy was actually happening. The Penobscot River Restoration Trust—a group including American Rivers, the Atlantic Salmon Federation, Maine Audubon, the Natural Resources Council of Maine, the Nature Conservancy, Trout Unlimited, and the Penobscot Indian Nation—had struck a deal with power companies and the state and federal government. Those painful dam battles of the 1980s and 1990s left a legacy, leading the power companies to consider selling their dams rather than fighting to license them again in coming decades. The trust had raised some $63 million and bought three dams on the Penobscot; one had already been dismantled, and the Veazie dam was scheduled to come out in July.

Excited biologists and environmentalists and anglers were talking about the Penobscot as an opportunity to see what happens when a river comes back to life. And though the Penobscot's dams aren't the only ones to come down in recent years—the Edwards Dam on Maine's Kennebec River was breached in 1999, and the ongoing dismantling of two towering dams on the Elwha River in Washington State is physically the largest such project in US history—the scope of the Penobscot restoration, which could impact more than 1,000 river and stream miles, is unique.

Even riverbeds shift, I realized. So in late July I found myself in a crowd gathered below the Veazie dam, watching backhoes mounted with pneumatic drills crawl onto its crest. Members of the trust's many partners were there on the riverbank, including a delegation from the Penobscot Nation. There were old-time fishermen and nature-lovers and quite a few curious locals: someone told me they looked forward to "facing the river," implying that for too long, people hadn't felt much connection to it.

I put one of Dad's old lenses on my camera and, as the drills punched a hole in the dam's crest, snapped a photograph. As the water spouted, the crowd cheered. I joined in, too, despite some mixed emotions. I was happy on Dad's behalf, but the whole thing made me wonder: Would the salmon come back? Could the Penobscot truly be restored? And in the context of a watershed that has been dammed up for centuries, what did restoration even really mean?

Rory Saunders was a high school classmate of mine, though we didn't hang out much together or stay in touch afterward. If we had, I might have learned that he spent his boyhood clambering the banks of the Kenduskeag, a stream that flows behind my mother's house, where I still walk almost every day when I'm home for a visit. (In the Penobscot Indian language, Kenduskeag means "eel-weir place," one of many spots on the river with names recalling the historical abundance of fish.) Saunders went to Wisconsin for graduate studies on smallmouth bass, becoming a marine biologist, then returned to Maine in 2002 for his dream job: helping restore the Penobscot.

To understand river ecology, Saunders took me to a stream called the Sedgeunkedunk—"rapids at the mouth"—which tumbles 3 miles from a small lake and its surrounding wetlands into the Penobscot, a few miles below the Veazie dam. Several years ago, two old dams were removed from this tributary: one at the Sedgeunkedunk's mouth and another below the wetlands. The latter dam was replaced by a long, rocky ramp that allows fish to pass over it while maintaining water elevation in the wetland above. To Saunders and other researchers, this new access made Sedgeunkedunk a living preview of what might happen throughout the Penobscot basin as the big dams came down.

Steve Coghlan, a University of Maine biologist, joined us at the stream. "It's great that the dams are coming out of the main river, but to me it's not very interesting," he said. "Really the main stem river is a highway for fish to get to places like this." Through Coghlan's eyes, the relatively wide, slow Penobscot, running 109 miles inland from the sea, is merely a prelude to thousands of miles of streams and creeks, hundreds of lakes and ponds and wetlands. These, he explained, are systems for turning light and water into nutrients and life.

Sediments in the lakes and streams nourish microbes and algae, which in turn feed insects that are preyed on by smaller fish, then larger fish, then the creatures that feed on them: otters and osprey, cormorants and turtles and songbirds. More species, each with different life histories and habits, mean that more nutrients are processed. Post-dam, Coghlan said, the Sedgeunkedunk is a more biodiverse place, a far richer place, than before.

Atlantic salmon patrol the stream now, as do sea lamprey—much-maligned, eel-like fish that feed by attaching themselves to other fish, boring holes in their sides. Coghlan came to Maine from upstate New York, where it wasn't uncommon for biologists to poison entire streams in hopes of exterminating lamprey. At Sedgeunkedunk, he has found them to be, unexpectedly, a keystone species: to build spawning nests they thrash rocks into place, in the process dislodging fish eggs and invertebrates for other creatures to eat, and loosening gravel for salmon to build their nests. When lamprey die, their decomposing bodies provide a burst of food for insects and microorganisms at the food chain's base.

Downstream from the lampreys' carcasses, Coghlan has discovered, biological productivity explodes. That productivity may extend onto land: alewives and lamprey fattened at sea may once have constituted a biomass comparable to the northwest Pacific salmon runs, which are thought to have fertilized the region's great forests. Transported by birds and riverine mammals into the surrounding landscape, alewives and lamprey may have likewise nourished eastern Maine's pre-industrial forests. Nutrients went downstream too, of course, in the form of out-migrating juvenile herring and salmon. In turn, those fish populations would have supported striped bass and mackerel, haddock, tuna, and cod, the foundations of New England's commercial fisheries.

Fish swimming from a thousand Sedgeunkedunks were, to echo Aldo Leopold's elegy for passenger pigeons, the lightning that passed between the Atlantic Ocean and Maine's interior, cycling nutrients and energy, spawning a plenitude of life. As the big dams come down, that plenitude has a chance to multiply again. But Coghlan and Saunders raised an important caveat: there are still more than a hundred small, old dams, like the ones removed from Sedgeunkedunk, along the tributary streams. For a true restoration to occur, they'll need to come out, too—maybe not every one, but at least those with the most strategic importance. "Imagine the Penobscot is a big jigsaw puzzle," said Coghlan. "There's all those gaping holes. We're putting the pieces back."

Dad's Penobscot Salmon Club is shuttered now, but a pulse remains upstream at the Veazie Salmon Club, where on a rainy morning in July five men gathered to play their daily cribbage game. Joe King, one of the club's founders, sat at the table. Now eighty years old, King can tell stories about what it was like to catch salmon in the 1940s, a boy with a bamboo pole and a family to help feed. Thirty years ago, he said, the club would have been standing-room-only, with anglers waiting out the rain. Now the only crowd can be found in photographs lining the walls.

King thinks the dam removal and river restoration are a terrible idea, a $63 million boondoggle doomed to fail. It's not that he doesn't want to see the river come back—he just doesn't think it's possible. Part of this may reflect a lingering bitterness; salmon fishing on the Penobscot was made illegal in 2009 when the fish were declared federally endangered. He doesn't trust biologists and environmentalists, who have sometimes ignored and condescended to the Penobscot's anglers.

In some ways, King's pessimism about salmon restoration overlaps with emerging research. Warming waters in the northwest Atlantic, where salmon feed, have shifted food chains; capelin, the salmon's preferred food, are now smaller and less nourishing. They may no longer be able to support historically large salmon populations. Yet even smaller numbers would still be welcome, given the salmon's current endangered state, and whatever chance salmon have will only improve with better access to the Penobscot and their upstream spawning grounds.

On one matter, though, King's doubts ring true: The restoration's success hinges largely on a device called a "fish elevator," which is about as strange a concept as it sounds like, on the one remaining lower-river dam in Milford, about 15 miles north of Bangor.

This is the asterisk to the Penobscot's restoration: Though two dams have been removed, and a third will be shut down and circumvented by an artificial stream next year, the Milford dam will remain in operation and generate more power than before. That was the deal the utilities and government and environmental groups struck. It wasn't ideal, at least from an environmental perspective, but it was the only way to get the other dams out. In return, to give the restoration a fighting chance, the Milford dam's owner, Black Bear Power, will install a fish elevator—a system that, at least in theory, will carry migrating salmon and shad and all the other upstream species successfully past the dam.

"That was one of the selling points: this is going to be state-of-the-art," King says. "There's no such thing." And indeed, the track record of fish elevators is generally bad. As John Waldman and colleagues noted earlier this year in a *Conservation Letters* paper, most major-river dams in the eastern United States already have elevators or ladders, which are supposed to allow fish to climb upstream. If they worked, the populations wouldn't be so ghostly.

For species that don't go far upstream—including tomcod, sturgeon, striped bass, and rainbow smelt—the elevator is not a worry. The dams closest to the sea have been removed, and historically these fish didn't swim beyond Milford, where the river's elevation drops sharply. But for fish that travel far upriver, the elevator is a concern. Each species has its own strategy, evolved over millions of years, for ascending rivers. An elevator constricts those strategies into a single, unfamiliar approach that all fish must make through a single concrete channel into a van-sized lift chamber.

Even finding the elevator's door can be a challenge. In concept, migrating fish will sense a current diverted into a passage to the elevator and follow it. But they must distinguish that signal from regular flows of the river, including snowmelt-swollen water spilling over the dam. Later in the year, fish going downstream must find the passage without mistakenly swimming over the dam's lip and onto a concrete apron below, or being sucked into turbines that can just as easily generate fishmeal as they can electricity.

I asked Saunders, my old high school classmate turned National Oceanic and Atmospheric Administration (NOAA) biologist, about the poor track record of fish elevators. He acknowledges that getting it right is a major challenge, but doesn't concede defeat. Fish passages aren't intrinsically flawed, he says. Scientists just need to figure out how to make them work. Too often, the blueprints for Eastern elevators have been copied from systems built to transport Pacific salmon rather than designed with local species in mind. They also need to be run correctly: when the Milford elevator becomes operational next spring, someone from Black Bear Power will need to monitor it constantly, sending fish upstream when it's full and making sure the doors aren't closed when they approach.

Black Bear Power has pledged to pass 95 percent of migrating salmon safely through the dam, but that's a nonbinding number, and there are no requirements for other species. If the elevator doesn't work as advertised, improvements might be made—but what that means, and who will foot the bill, is the sort of fine-print detail that will become clear only in years to come, and could eventually loom large.

All that, though, is something to consider in the future. Saunders doesn't ignore the importance of these challenges, but he also puts them in context. A few decades ago, when Dad fought so hard just to prevent new dams from being built, when the big dams seemed permanent, challenges like these were practically inconceivable. They'd have felt like an opportunity. Fixing a fish elevator should be far easier than removing the dams. At least it's now possible to imagine the river coming back to life.

"I never thought this day would come," Saunders says, echoing my father's sentiments. "This is a chance to explore whether modern civilization and migratory fish can coexist. It's our best chance at getting a river right."

Author's note: The dam removals have been a resounding success. In 2016, nearly 2 million river herring returned to the Penobscot—a full 1,000 times more herring than swam upstream five years earlier, before the removals. Shad populations have swelled from a few dozen into the thousands; and though salmon numbers remain low, they're rising steadily. "Fish biomass is going through the roof," Saunders told me. The Penobscot is alive again.

A CHIMP'S DAY IN COURT: INSIDE THE HISTORIC DEMAND FOR NONHUMAN RIGHTS

On the morning of December 2, 2013, a lawyer named Steve Wise and three other members of the Nonhuman Rights Project walked up the steps of the Fulton County Courthouse in Johnstown, New York, and into the annals of history.

In Wise's hands was a lawsuit demanding something unprecedented in American law: formal recognition that a twenty-six-year-old chimpanzee named Tommy, kept alone in a cage in a local warehouse, is a person, possessing a legal right to bodily liberty previously reserved for humans.

One day later, Wise delivered a similar lawsuit in Niagara Falls, on behalf of Kiko, a chimpanzee living in the home of a couple who purchased him twenty years ago from an Ohio ice cream parlor. Two days after that, Wise filed suit on Long Island, claiming freedom for Hercules and Leo, two chimps used in locomotion studies at Stony Brook University and owned by the New Iberia Research Center (NIRC), a facility infamous for the abuse of its animals.

What Wise and the members of his Nonhuman Rights Project assert is both simple and radical: that personhood—the legal prerequisite for having rights of any sort at all—should no more be based on species classification than it is on skin color. Instead, the lawsuits argue that personhood derives from cognitive and emotional qualities that chimpanzees, like humans, possess in abundance.

Tommy and Kiko and Hercules and Leo are not humans, acknowledges Wise, deserving full human rights. They shouldn't have a right to vote, or to enter a restaurant or a public school. But they and every other captive chimpanzee are enough like us to have a right not to be owned and imprisoned against their will. If it's not realistic to release them into a wild they've never known, they might be sent to chimpanzee sanctuaries, free to live with a modicum of the liberty they deserve. Not because we've deigned to be nice to them, but because it's their right.

Whether a judge will agree remains to be seen, and most legal scholars, even those who support the case, think it's unlikely. Never in the history of Western law have ideals of liberty and equality been formally applied to an animal other than *Homo sapiens*. Judges don't like being radical. The lawsuit also raises what many consider an unacceptably impractical possibility: personhood demands filed on behalf of not only chimpanzees but also animals used in research and on farms, or living in our landscapes.

Legal implications aside, however, the argument for chimpanzee personhood draws on decades of scientific research on humanity's closest living relative. And where the argument turns from law to science, the Nonhuman Rights Project's claims are backed with affidavits filed by nine of the world's leading primate researchers, describing in exhaustive detail the scientific evidence for regarding Tommy, Kiko, Hercules, and Leo as extraordinarily complex, self-aware, and autonomous creatures comparable in many ways to ourselves.

Such information has previously been applied to questions of welfare, to cage sizes and research justifications, but not to fundamental legal questions of personhood. Yet this is where the science points, and though many people dismiss these claims as quixotic and impractical, Wise and his compatriots are following scientific fact to an arguably logical legal conclusion.

It's still entirely possible that the lawsuits will be dismissed without Wise opening his mouth in a New York courtroom. But it's also possible that a judge will at least agree to hear their case. If and when that happens, a chimpanzee will finally have his day in court. "It doesn't matter if you're a human or a chimp," said Wise. "If you have the cognitive capacity to live life as you choose, you should be able to do that. Your species is completely, completely irrelevant."

The lawsuit is the culmination of a legal strategy seeded three decades ago when Wise was practicing animal protection—or, as he prefers to say, animal slave—law in Massachusetts. He litigated hundreds of cases, ranging from the unusual mistreatment of dogs and cats to more unusual lawsuits: attempts to stop deer hunts, the transfer of a dolphin from the New England Aquarium into sonar research with the US Navy.

Sometimes, as when an especially egregious abuse violated local animal cruelty statutes, Wise would win. But often his lawsuits were dismissed by judges who argued that Wise couldn't sue on behalf of animals he or his clients didn't personally own. In legal terms, they lacked standing: the ability to show how humans would suffer from the mistreatment of an animal.

It might sound illogical that an animal's suffering could be legally irrelevant unless it directly affects a human, but "these are the kinds of gymnastics that animal protection lawyers have to go through," said Wise during a September talk at George Washington University. "Nonhuman animals are legal things, and they have always been legal things. That means they're essentially invisible to our civil law."

Cruelty laws aside, some animal protections do exist, such as those for endangered species. Yet even those reflect the value society places on populations and species units. They say nothing about the personhood of an individual animal. And though individuality might be acknowledged in, say, requirements that research chimpanzees be given toys to play with, they reflect token generosities rather than any acknowledgement of an animal's essential personhood.

"The example I give is that I can take a baseball bat, and smash the window of your car, and I'll be charged with something," said Wise during the talk. "But the car or the windshield doesn't have legal rights. It's not a person. Essentially, a nonhuman animal is a kind of animate windshield. If I'm cruel to her, I can go to jail, but the nonhuman animal is a complete bystander to this. She doesn't have any rights."

Wise furthered the analogy with references to animals as legally equivalent to tables and chairs—language that, to anyone who knows a cat or dog, or really has any experience with animals whatsoever, will sound strange: animals are self-evidently not these things. Yet as Wise argues in *Drawing the Line*, a 2002 book in which he articulated the legal and scientific framework of the lawsuits, Western law's view of animals has been shaped by an intellectual tradition that might charitably be called ignorance, from Aristotle's proclamation that, just as slaves were made to serve their masters, so animals were made to serve humans, on through René Descartes' characterization of animals as clockwork automata.

Whether this intellectual history is directly responsible for modern law's treatment of animals is a matter of debate for legal historians. The fact remains, though, that Tommy—whom Wise visited personally in October, under the guise of being curious about a herd of holiday-rental reindeer also kept at a trailer dealership run by Tommy's owner, and whom Wise found alone in a newspaper-lined cement cage in a dank warehouse with nothing to occupy his attention but a small color television—might as well be a windshield.

"There is no question this court would release Tommy if he were a human being," read the documents. "There can be equally no doubt that Tommy is imprisoned for a single reason: despite his capacities for autonomy, self-determination, self-awareness, and dozens of other allied and connected extraordinarily complex cognitive abilities, he is a chimpanzee."

Making a factual case for those capacities are the affidavits filed by nine scientists whose expertise spans not only primatology but also psychology, neuroscience, animal behavior, anthropology, cognition, and learning theory. The scientists are not themselves members of the Nonhuman Rights Project, and the affidavits don't talk about law, only science. Gathered together, they amount to more than two hundred pages of text and references, reviewing hundreds of studies—not just their own but from hundreds of other researchers—on dozens of cognitive, neurological, and behavioral traits.

Lawsuits aside, the affidavits are a veritable textbook of contemporary and historical chimpanzee research, and if conveying their full detail requires more space than a journalistic article affords, a few themes stand out. At a purely neurological level, chimpanzees share many of the same features—enlarged frontal lobes, asymmetry between left- and right-brain functions, specialized cell types in specific brain regions—associated with high-level cognition in humans.

Mary Lee Jensvold, an experimental psychologist and former director of the Chimpanzee and Human Communication Institute, who worked for twenty-seven years with chimps taught sign language, describes the conceptual richness of their communicative abilities, which include symbolic thought, perspective-taking, and references to past and future events, and the parallel manner in which these develop in both human and chimpanzee infants. These are evidence, say she and others, of deep cognitive similarities.

Similar parallels can be seen in brain development, writes the primatologist Tetsuro Matsuzawa of Kyoto University, current president of the International Primatological Society. He also discusses their highly sophisticated numerical and sequencing abilities, which have been the subject of public fascination: chimps out-calculating human children and even college students! These math skills, explains Matsuzawa, indicate the sophistication of memory systems considered central to planning and making decisions rather than acting on instinct or automatic, stimulus-response conditioning.

Another series of well-known findings involve tool-making and cultural traditions, both of which were once thought to be uniquely human capacities. William McGrew, an evolutionary primatologist at the University of Cambridge, describes how chimpanzees learn from one another to make complex tools that resemble those used by Stone Age humans and even some surviving aboriginal groups. "The foraging kits of some chimpanzee populations, such as in western Tanzania, are indistinguishable in complexity from the tool kits of some of the simplest material cultures of humans, such as Tasmanian aborigines," McGrew writes.

Different methods of tool manufacture and use are considered hallmarks of distinct chimpanzee cultures, of which more than forty exist in Africa. Cultures are also distinguished by social traditions: McGrew recounts how the famed primatologist Jane Goodall, who sits on the Nonhuman Rights Project's Board of Directors, documented males of a group she studies performing slow, rhythmic dances at the beginning of rain seasons. It's not unreasonable, Goodall says, to think these might be rain dances.

Whether that's so is speculation, but the cognitive faculties underlying culture are not. In chimpanzees as in humans, writes McGrew, social learning requires emulation, imitation, and an ability to explain, all of which signify a sense of self. That's hardly not unique in the animal kingdom—even a cockroach likely has self-awareness—but there's reason to think that a chimpanzee's self-awareness is comparable to our own. Not only can chimpanzees recognize themselves in mirrors, which is considered a sign of sophisticated self-awareness, but they also recognize photographs of themselves as youngsters.

Perhaps crucially, they're capable of "mental time travel," said Matthias Osvath, a cognitive zoologist at Sweden's University of Lund. Known for his research on Santino, a chimpanzee who makes elaborate plans to throw stones at zoo visitors, Osvath describes mental time travel as the ability to imagine oneself in the past and future as well as the present. This capacity underlies an essential aspect of human experience, Osvath said: we engage in some form of it whenever our own thoughts stray from the immediate moment and when we make plans.

It's this ability that produces the autobiographical, first-person sense we have of our own lives, and it could very well mean that Tommy is both aware of the misery of his situation and can contemplate a future of never-ending captivity. If so, that would not in itself be the grounds for the Nonhuman Rights Project's personhood claim; rather, it would be a facet of what the various capacities of chimpanzees, from complex memory and self-awareness to imagination and mental time travel, signify: autonomy.

As best as we can tell, Tommy and Kiko and Hercules and Leo don't live their lives according to a predetermined program. They can go to sleep at night thinking of what they might do the next day, then wake up and do it—or not. They live consciously, with full awareness and exercised choice.

This assertion, that the cognitive qualities enumerated amount to autonomy, is arguably a judgment that drifts away from established fact and toward a philosophical argument about the nature of autonomy. But as the psychologist James King of the University of Arizona, a former editor of the *Journal of Comparative Psychology*, notes in his affidavit, autonomy isn't something we can directly observe even in humans.

Rather, we judge it by behavior, and "evidence for autonomous behavior in humans is not seriously disputed," writes King. "In chimpanzees, the behavioral evidence for autonomy is not seriously disputed." And if we can't measure autonomy the way we can, say, gravity, it's the simplest explanation for what we can observe in chimpanzees.

A few decades ago, before most of this research was conducted, when editors at the journal *Nature* castigated Jane Goodall for referring to chimpanzees with gender-specific pronouns, such claims would have been deeply controversial. After all, noted Osvath, science too had to free itself from the intellectual legacy of Aristotle and animals-as-machines. But the essential outlines are widely accepted.

To be sure, said Brian Hare, a Duke University evolutionary anthropologist who has studied cognition in chimpanzees and dogs, individual pieces of the cited research might still be debated, but over fine points of similarity and difference. "If you're talking about it in a quick way, all those statements are accurate, but in each of those categories where chimpanzees might be similar to us, there's different types of equivalence," Hare said.

In other words, a chimpanzee has an autobiographical sense of self. Does he spend as much time as we do contemplating the future? Does he experience autonomy the way we do? That we don't know for sure, but the Nonhuman Rights Project argument isn't for one-to-one equivalence, but rather substantive similarity. Indeed, that's essentially the judgment passed by an Institute of Medicine committee convened in 2011 to advise the National Institutes of Health on its use of chimpanzees in research: they're just too much like humans to use in harmful experiments unless there's an overwhelming, human-life-saving reason to do so.

To all this, one might raise the question: Could a chimpanzee ask to be free? In the wild, where chimpanzees are known to communicate using abstract, symbolic gestures, that might indeed be possible, but Mary Lee Jensvold recalled a more germane story. It involves chimpanzees trained to use sign language by Roger Fouts, the Chimpanzee and Human Communication Institute's founder, early in his career, when Fouts was still at the University of Oklahoma's Institute for Primate Studies.

The chimps were sold to the Laboratory for Experimental Medicine and Surgery in Primates in New York. Years later, one of Fouts's collaborators, the Gonzaga University comparative psychologist Mark Bodamer, visited the laboratory. "He met the signing chimps," said Jensvold. "The chimps were signing. One of the technicians said, 'They're always doing that.' And one of the chimps signed, 'Key. Out.'"

If the Nonhuman Rights Project's story were a movie, it would cut now to a montage of hotel conference rooms, take-out cartons, and ever-growing piles of paper strewn around the project's principals. There's Wise, of course, his hair grayer and face more lined than in the dust-jacket photos of his books; Elizabeth Stein, Wise's intellectual sparring partner, who left financial law to defend animals; Monica Miller, a recent law school graduate who handles their legal database dives and archive searches; the executive director Natalie Prosin, who does a little of everything; and, appearing via Skype on a laptop, Lori Marino, the Emory University neuroscientist who first showed that dolphins could recognize themselves in mirrors.

Over the past two years, with volunteer help from dozens of animal law students and lawyers, the group analyzed the laws of all fifty states, plumbing the arcana of nineteenth-century slave cases and legal guardianship precedents, posing for each some sixty fine-print questions designed to identify the most legally hospitable state for their lawsuit. Earlier this year they picked New York, where courts have historically been generous in their willingness to hear personhood claims, and decided to bring suit on behalf of every chimp they could find.

Two of these, named Merlin and Reba, kept in a cage at a zoo in the upstate town of Catskills, died before the case was ready. Merlin was the second to die, three months after Reba, from complications resulting from an abscessed tooth. According to the Nonhuman Rights Project, he spent the last several weeks of his life punching himself in the face. A third chimpanzee, a twenty-eight-year-old named Charlie the Karate Chimp, who was owned by the same couple who own Kiko, died in early November. The couple, Carmen and Christi Presti, run a self-described primate sanctuary previously known as Monkey Business; they'd purchased Charlie as a fifteen-month-old, later teaching him karate and displaying him on TV shows including *America's Funniest Home Videos* and *How I Met Your Mother*.

The Prestis did not respond to requests to be interviewed, nor did Tommy's owner, Patrick Lavery, or Stony Brook University, where Hercules and Leo are used in the anatomist Susan Larson's research on primate locomotion. Both are owned by the NIRC, to which a third Stony Brook chimpanzee, named Carter, was returned last August. A chimpanzee named Carter is also mentioned in a 2007 Humane Society undercover investigation at NIRC, which has been accused of illegally breeding chimps.

The HSUS allegations described several hundred NIRC chimpanzees living in barren, windowless concrete rooms, with literally nothing to do, sometimes in isolation and at other times packed into cages. Infants were routinely taken from their mothers shortly after birth; in the wild, they'd have nursed for five years and remained with their mothers for several years after that. At NIRC, juveniles often displayed the compulsive rocking typical of abandoned human children. "Danielle gave birth to son Carter on 9/1/06," read the report. "Carter was taken from his mother so she could be bred again."

New Iberia was ultimately fined a pittance of $18,000 for violating the Animal Welfare Act. This sort of treatment, and the institution of chimpanzee ownership that supports it, isn't just wrong, contends Wise. It violates their rights. To make that argument, he's appealing to common law, a body of jurisprudence distinct from constitutional law—Wise doesn't argue, and has openly criticized the notion, that the Thirteenth Amendment should apply to nonhumans—and from statutory law, which is passed by legislatures. Common law is where everything else gets settled, and at least historically, it's where judges have the latitude to make and interpret law in keeping with changing social standards and scientific knowledge.

Personal injuries fall under the auspices of common law. So do rights to privacy and interfamily disputes. It was also in a common-law court where a slave's personhood was legally recognized for the first time. That case, *Somerset v. Steuart*, took place in England in 1772, and is the subject of a book by Wise, *Though the Heavens May Fall*, in which he details the legal arguments that convinced one of history's most influential jurists, Lord Mansfield, himself a slave owner, that a runaway slave named James Somerset was a person with a right to be free.

That case and its arguments, including the legal strategy of filing a writ of habeas corpus, an ancient legal principle—it translates from Latin as "show the body"—used to challenge unlawful imprisonment, was formative for Wise. It guides the Nonhuman Rights Project's lawsuits, which ask for writs of habeas corpus on behalf of the chimpanzees, and is among the dozens of cases and legal theories cited in its court documents.

These run from *Somerset v. Steuart* and *The Collected Works of Abraham Lincoln* to tomes of Roman civil law and *Lemmon v. People*, an 1852 case in which the New York Supreme Court allowed a dock worker to file habeas corpus writs on behalf of eight slaves held on a boat in New York harbor. That precedent, hopes Wise, will convince a judge that the Nonhuman Rights Project can represent the chimpanzees. They also quote John Chipman Gray, an influential late-nineteenth- and early twentieth-century legal scholar, who said that "personhood as a legal concept arises not from the humanity of the subject but from the ascriptions of rights and duties to the subject."

That is, in a formal nutshell, the essence of the Nonhuman Rights Project's argument: a person—an entity capable of possessing legal rights—does not need to be human. To a certain extent this has already been demonstrated by the legal personhood afforded to corporations and ships, but those are still human institutions. Wise is calling for personhood based purely on principle.

"Legal person has never been a synonym for 'human being,'" reads the lawsuit. "A free being such as Tommy, who possesses autonomy, self-determination, self-awareness, and the ability to choose how to live his life, must be recognized as a common law 'person.'" The rights to which Tommy is entailed, they stress, don't include nearly all the rights given to humans—he's not entitled to a free education or emergency health care—but by virtue of his autonomy, his capacity for self-determination, has a right not to be imprisoned.

This sort of right is technically known as a "dignity right," and Erin Daly, a Widener University School of Law professor who has written extensively about dignity rights, called their proposed grounding in autonomy "a reasonable and interesting argument," though unprecedented. "I don't know of anything that says, 'Anything with autonomy has dignity rights,'" she said—not because the idea has been rejected but because nobody's yet made that case so explicitly.

"As a legal culture, we do care about autonomy. It's an important value in America, and more so in America than in other countries," Daly said. "I think that makes sense to test whether nonhumans should be accorded dignity rights in some way."

Should a judge feel uncomfortable with the idea that self-determination is an intrinsic basis for liberty, the lawsuit offers a related but subtly different argument, grounded instead in equality: the principle by which, if one person is similar to another, they are treated equally by the law. It's on this principle that rights for blacks, other racial minorities, women, and nonheterosexuals have been granted—and not because they were necessarily considered the full equals of existing rights-holders, but because they were equal enough.

Likewise, even humans whose cognitive capacities are well below those of chimpanzees—people with severe mental disabilities, infants whose brains won't ever develop, adults with severe dementia—are granted some rights. They might be kept in institutions, or have certain decisions made for them, but they can't be owned or experimented on. That isn't simply because they satisfy some arbitrary genetic standard, argues Wise, or qualify via legal loophole for the protections of fully abled humans. It's because they have capacities of autonomy and selfhood, however limited, that we as a society value so deeply.

"We'll go in front of a judge and say, "Whatever is the reason why a human being should have a right to bodily liberty, whatever trait they have, whatever characteristic they have, our chimpanzee has it too," Wise said. "You can name whatever you want. Our chimp has it too."

"The strength of Wise's case is that he's called the law into question on its fundamental values of dignity, equality, and autonomy," said Paul Waldau, who teaches animal law and ethics at Harvard Law School. "He's saying, 'If you will just accept these as the really beautiful foundations of our law, then I win.' He's using the real foundational arguments of the legal system to do this. And in a way it's quite conservative: He's saying, 'The basic values are good—and watch how, if you extrapolate them carefully, they will take us: to legal rights for chimpanzees.' That's beautiful. It affirms the system, even as it asks the system to go beyond what it presently has done."

Beautiful as the argument might be, however, can it succeed? Even among legal scholars who respect Wise's arguments and acknowledge changing human relationships to animals—widespread condemnations of cruelty, insistence on humanely treated farm animals, the NIH's historic rejection of most chimp experimentation, a fifteen-fold increase in academic animal-law programs over the last two decades—it's difficult to find one who thinks a judge will rule in the chimpanzees' favor.

"I think Wise's work as academic theory, as policy, as discussion, is second to none," said Jonathan Lovvorn, a Georgetown University law professor and senior litigator at the Humane Society of the United States. "His work was transformative within the legal academic field. But I think that translating it at this stage into a legal strategy is just not feasible."

The problem, said Lovvorn, is that chimpanzee personhood, however limited, is still too radical for most judges to consider. Though articulated in bedrock principles of liberty and equality, it's still about an animal. A ruling in the chimpanzees' favor would represent a legal sea change; in the profession's argot, it's "too big an ask," said Lovvorn. It's true that common law has historically been flexible, a vehicle for judicial activism, he said, but that was far more true in the eighteenth and nineteenth century than it is now.

Other reservations that will almost certainly cross a judge's mind were raised more than a decade ago when Wise debated Richard Epstein, a legal scholar at New York University, and Richard Posner, a judge on the 7th US Circuit Court of Appeals. "Chimpanzees and bonobos may be the obvious candidates," said Epstein, but that could open the door to personhood claims on behalf of cattle or sheep or rats and mice used in human life-saving medical research. "It's not clear how far exactly you want to run this," he said.

To this, Wise would rejoin that the lawsuit is about the possibility of granting a single right to chimpanzees, and has nothing to do with sheep or rats—but it's no secret that it represents what Wise has often described as his goal of breaking down the legal wall that now separates humans from all other animals. "Once we begin to punch through that wall," said Wise, "all sorts of roads possibly open."

Chimpanzees are now considered exceptional, but similar arguments could presently be made on behalf of other great apes, elephants, and some cetaceans, which the Nonhuman Rights Project had also considered. Given that many scientists now recognize consciousness across the animal kingdom, autonomy may eventually prove to be an easily attainable standard for personhood.

One might argue that it's only ethically and scientifically consistent to judge the personhood of each species empirically, as the Nonhuman Rights Project does with chimpanzees, and contemplate rights on a species-by-species basis. From a certain perspective, this is even exciting: If nonhuman animals can be persons, then what sort of persons should they be? What rights should they have, or not? Might an entirely new class of rights possession, "animalhood," be established? But excitement is not what most career- and reputation-minded judges seek.

Moreover, as Posner noted, Wise is likely to encounter a deep-seated, almost primal objection that legal rights are simply human rights, however arbitrary that may seem. And as Posner explained in *How Judges Think*, judges often go with their gut. Of the many factors guiding their decisions, wrote Posner, what the law actually says is often the least important.

In the meantime, says Lovvorn, there's a danger that the Nonhuman Rights Project's efforts will distract from more practical, immediately realistic animal welfare reforms, such as improvements in farm animal treatment or reductions in biomedical experiments on animals. Opponents of such reforms, said Lovvorn, might seize on chimpanzee personhood as an example of the radical changes that ostensibly modest reformers really want. "I can't tell you how many times we hear people say, 'We can't give pigs or laying hens enough room to turn around and stretch their wings because the end game here is personhood for animals and an end to their commercial use.'"

Brian Hare also struck a practical tone, noting that reforms already underway to protect chimpanzees, such as eliminating a federal loophole that grants endangered status to wild but not captive chimps, could accomplish some of the Nonhuman Rights Project's goals while avoiding thorny questions of personhood. "I think it's great that he's asking the question and challenging the law," said Hare, "but I'm not sure that this particular solution is necessary, or in the best interest of the chimpanzees themselves. I think they are fighting for rights when they should be fighting for welfare and conservation."

William McGrew, however, expressed reservations about the limits of good intentions in the absence of enforceable rights. McGrew said he used to think that "once humans knew of the apes' qualities, people would extend their concern to their closest living relations." Now, "being older and wiser, I see that was unrealistically idealistic. Just like minors and minorities amongst humans, chimpanzees need legal protection"—and that, McGrew said, should include recognized rights of life, liberty, and the pursuit of happiness.

Relying on endangered species protections also raises the question of whether chimpanzees would remain protected if someday they're not endangered. Would they deserve any less consideration then? Rights, argues Wise, are a matter of principle, not something to abandon when they seem inconvenient. Neither are they uncontroversial or unthreatening, at least not when they're first granted. And even if courts rule against chimpanzee personhood—something that, Wise concedes, is quite possible—at the very least a judge might have the courage and integrity to hold the discussion openly, in light of scientific fact and legal principle.

That discussion could well begin in coming weeks, as judges consider the lawsuits now before them. They might dismiss the cases summarily, without a hearing; or, if one decides to issue the habeas corpus writ, then the chimpanzee's owner will be ordered to appear in court and justify the animal's detention. Wise hopes that, even if judges aren't prepared to accept that a chimpanzee could have a right, they'll be courageous and open-minded enough to listen to the arguments.

If the Nonhuman Rights Project loses, said Wise, they'll appeal the decision; if they lose the appeal, they'll review their arguments and try again, and again after that. Yet there's at least a possibility, however slim, that a judge will rule in their favor, and make possible the lawsuit's concluding request: That Tommy, who can't be released to Africa, can be delivered from solitary confinement in a warehouse cage and taken to a sanctuary, "there to spend the rest of his life living like a chimpanzee, amongst chimpanzee friends, climbing, playing, socializing, feeling the sun, and seeing the sky."

CHIMPANZEE RIGHTS GET A DAY IN COURT

More than a year after the starting fight for legal personhood for the research chimpanzees Hercules and Leo, the apes and their lawyers got their day in court. At a hearing in Manhattan on Wednesday, a judge heard arguments in the landmark lawsuit against Stony Brook University, with a decision expected later this summer. At stake: the question of whether only human beings deserve human rights.

A decision could set a precedent for challenging, under human law, the captivity of other chimpanzees—and perhaps other species. It's a radical notion, and many legal experts doubted whether the lawsuit, one of several filed late in 2013 by the Nonhuman Rights Project, would ever reach court.

But Justice Barbara Jaffe decided to consider the arguments. "The law evolves according to new discoveries and social mores," she said while presiding over the hearing. "Isn't it incumbent on judiciaries to at least consider whether a class of beings may be granted a right?"

Jaffe posed that question to the New York assistant attorney general Christopher Coulston, who represented the university, where the two chimps are housed. Coulston had argued that Jaffe was bound by the previous decisions of two appellate courts, which had ruled that other Nonhuman Rights Project chimps didn't qualify for habeas corpus, the legal principle that protects people from illegal imprisonment.

Both those decisions are controversial. In one, judges decided that habeas corpus didn't apply because the chimp would be transferred from one form of captivity to another—in this case, a sanctuary. But illegally held human prisoners have been released to mental hospitals, and juveniles into the care of guardians.

In the other appeals court decision, judges declared that chimps are not legal persons because they can't fulfill duties to human society. But that rationale arguably denies personhood to young children and mentally incapacitated individuals, as several high-profile legal scholars, including the constitutional law expert Laurence Tribe, pointed out. He filed a brief on behalf of the Nonhuman Rights Project, saying the court "reached its conclusion on the basis of a fundamentally flawed definition of legal personhood."

In fact, the Nonhuman Rights Project attorney Steven Wise argued, New York law requires judges to follow appeals court decisions involving only settled legal principles—which animal personhood is not. That set the stage for the pivotal question: What is the basis of legal personhood? Wise said it's rooted in the tremendous value placed by American society and New York law on liberty, which is synonymous with autonomy. "The purpose of the writ of habeas corpus isn't to protect a human being," he said. "It's to protect autonomy."

By that standard, Wise said, chimpanzees qualify. "Chimps are autonomous and self-determined beings. They are not governed by instinct," he said. "They are self-conscious. They have language, they have mathematics, they have material and social culture. They are the kinds of beings who can remember the past and plan for the future." In a human, argued Wise, those capacities are grounds for the right to be free.

Coulston marshaled an argument elsewhere made by Richard Posner, a legal theorist and federal appeals court judge who has written that legal rights and personhood were designed with only humans in mind. "Those rights evolved in relation to human interests," Coulston said. "I worry about the diminishment of those rights in some way if we expand them beyond human beings."

The cognitive capacities of chimpanzees have been compared to five-year-old humans, said Coulston; how would the legal system handle animals with minds comparable to a three-year-old or a one-year-old? "This becomes a question of where we're going," he said, with chimp personhood opening the floodgates to lawsuits on behalf of animals in zoos or on farms, or even pets. "The great writ is for human beings," he said, "and I think it should stay there."

Wise countered by saying that denying freedom to an autonomous being is itself a diminishment; it could even come back to bite us, serving as rationale for limiting human freedom. He described the slippery slope as a separate issue. Freedom—or at least sanctuary—for Hercules and Leo is something to debate on its own merits, just as rights for any potentially deserving human should be considered without regard for social inconvenience.

It is true, though, that success could lead to personhood claims on behalf of other chimps, as well as other great apes, orcas, and also elephants, for whom the Nonhuman Rights Project is now preparing a case. More than a third of Americans now support rights for animals.

Win or lose, Wise said at a press conference after the trial, the hearing itself was a victory. "Many human beings have these kinds of hearings," he said. Chimpanzees "are now being treated like all the other autonomous beings of this world." Whether they'll continue to get that treatment will be up to Justice Jaffe. Or—more likely—whoever hears the almost inevitable appeal of her decision.

Author's note: As of this manuscript's preparation, the Nonhuman Rights Project's claims had been denied in each court that heard them, as were their appeals. In many ways, though, they triumphed. Judges took their cases seriously; perhaps more important, so did the public.

The fate of the chimpanzee plaintiffs remains uncertain. According to the Nonhuman Rights Project, government records show that Tommy's owner sent him to a roadside zoo in Michigan in late 2015. The zoo, which has been criticized by animal advocates for mistreating another chimpanzee, reportedly denied this. As for Kiko, he remains with his owner, whereas Hercules and Leo were sent—after appearing in one last study, published in the journal *Nature Communications*—back to the infamous New Iberia Research Center (NIRC) in Louisiana.

In May of 2016, NIRC agreed to send its 220 chimpanzees to sanctuary. Among those chimpanzees, perhaps, is Carter, a chimpanzee reportedly sent to NIRC from Stony Brook University in 2012. I learned about Carter early in my reporting on the Nonhuman Rights Project, after stumbling on a flyer posted online by activists. Over the course of several years I tried without success to find any information about his health or whereabouts. He literally disappeared.

In my opinion, the arguments raised by Steven Wise and his colleagues are not about animal rights. They're a matter of civil rights.

MEDICAL EXPERIMENTATION ON CHIMPS IS NEARING AN END. BUT WHAT ABOUT MONKEYS?

The US government's decision to end its support of most medical experiments on chimpanzees came after decades of impassioned, often bitter debate—yet in some ways, it was an easy decision.

After all, chimpanzees and humans famously share 98 percent of our DNA. They're our closest living relatives. They're charismatic and enough like us that it's difficult not to treat them with compassion. But what about monkeys?

Even as the ethics of chimp research grabs headlines, the use of monkeys—tens of thousands in the United States alone—receives minimal attention. They arguably share many of the traits that make experiments on chimps and other great apes so ethically troubling, yet elicit barely a whisper of public concern.

That absence of attention says less about the ability of monkeys to think, feel, and suffer, than our willingness to think about it.

"It's been very hard to show differences between great apes and monkeys in terms of cognitive abilities," said the evolutionary anthropologist Brian Hare of Duke University, who has studied both. "I'm not saying there are not differences—of course there are—but it's hard to show. Most of what you know about great apes is also true about monkeys."

Nearly 120,000 nonhuman primates are kept in captivity in the United States, with roughly 70,000 used for research; more than

20,000 are imported every year, either sold as infants by monkey breeders or caught in the wild. Most are monkeys, and most of the monkeys are rhesus macaques, which have become the model primate of choice for medical research.

They're used to study dozens of diseases and conditions, from neurodegeneration and cancer to depression, diabetes, stroke recovery, and addiction. They're easier to work with than chimps, and solve what the geneticist Vincent Lynch of the University of Chicago calls a biomedical Goldilocks problem.

Monkeys are just different enough from humans "that we are comfortable experimenting on them and just close enough genetically that those experiments are still applicable to human health," Lynch said. "They would seem to be just right."

By "just right," Lynch doesn't mean that he works with monkeys himself. "They are just too similar to me to justify it to myself," he said. "Monkeys clearly are valuable as animal models, and I would never do experiments on them." It's the sort of sentiment often characterized as animal activist rhetoric, but it's also where the science is pointing.

Studies of rhesus monkeys have found them capable of empathy, long considered an essential human trait. They think about their own thoughts, which is essential for complex self-awareness. They can recognize themselves in mirrors, experience regret, have a sense of justice and fairness, and possess what cognitive scientists call theory of mind: an understanding of what other individuals think and feel.

Their brains possess anatomical features that, in humans, are central to emotion, and it makes intuitive sense that monkeys would feel deeply. After all, cognition and emotion are intertwined, and emotion is a deeply rooted evolutionary feature intertwined with living in large, social groups—which monkeys certainly do.

Monkeys share with chimpanzees nearly all of the features that a landmark Institute of Medicine report cited in concluding that chimpanzees are worthy of special consideration when assessing their use in research. Yet there's one crucial difference: unlike chimps, which are presently useful for studying just one or two diseases, monkeys are useful for many.

"Monkeys are especially valuable for a wide realm of research domains," said Stuart Zola, director of Yerkes National Primate Research Center. "Their close-to-human brain anatomy and close-to-human genome make them very good models," especially for brain-related diseases. Human and monkey immune systems "are organized in similar ways, so viruses and infectious agents can be effectively studied safely in monkeys."

Indeed, monkeys are only becoming more medically useful. The maturation of techniques for inserting disease-related human genes into monkeys "sets the stage for grand possibilities of clarifying the mechanisms of disease in ways heretofore never before accessible to us," Zola said.

How might this equation of monkey consciousness and medical utility be balanced? One possible response would be to say monkeys should not be used at all; another is to say that, however unfortunate, the possibility of easing human suffering outweighs any suffering in monkeys. Between these two positions is the argument that experiments on monkeys are acceptable, but only if the human benefits are truly significant.

For chimpanzees, the National Institutes of Health says they should be used in medical experiments only when no other alternatives exist. That high standard likely won't be reached in the near future with monkeys, but researchers might think more carefully about whether they want to use monkeys, as Vincent Lynch did. Institutional committees responsible for approving research proposals might also raise the bar for approving monkey experiments.

After the question of whether and when monkeys should be used comes the issue of how they are used. Tetsuro Matsuzawa, a Kyoto University primatologist and current president of the International Primatological Society, doesn't personally think monkeys should be used in medical experiments, but he also respects arguments in favor of using them to save human lives. Matsuzawa can accept monkey experiments, he said, if researchers treat them as compassionately as possible.

Hare echoed those words, saying that monkey welfare—even things as simple as improving the lighting in their housing—is often overlooked. "People are always talking about whether we should use them or not, and not about whether we can reduce their suffering," he said. "We need a culture of compassion, and we don't have that."

Federal laws and National Institutes of Health guidelines do technically require that researchers promote the welfare of their monkeys, but those rules aren't strictly enforced, said Hare. Only the most egregious violations are punished meaningfully, and researchers are given few incentives to improve their monkeys' lives.

"The question is, 'Have you demonstrated compassion for the animals in your own lab? Have you, on your own, without anyone patting your back, done something to improve how your monkeys are housed?'" Hare said. "It's a rare researcher who does that."

Kathleen Conlee, animal research director with the Humane Society of the United States, noted that the NIH has resisted attempts to strengthen the federal Animal Welfare Act, which guides how animals are used in research in the United States. Those rules are often vague and researchers can too easily apply for exemptions, said Conlee.

Many rhesus monkeys are, for example, kept in solitary cages, an experience known to drive these highly social animals crazy. "The Animal Welfare Act was supposed to create environments that addressed the psychological well-being of these animals," said Conlee. "A lot of the facilities are falling short of meeting those minimal standards."

One Humane Society analysis of documents from two major US primate research facilities found that their animals spent, on average, 53 percent of their lives in solitary housing, sometimes with nothing in their cage but a piece of metal hung on one side for "enrichment."

It might sound strange to consider how best to treat monkeys that are consigned to medical sacrifice anyway—but if we're going to use these intelligent, emotional creatures, we should do it right, said Conlee. "Until the day we ultimately replace them in research," she said, "we have an obligation to address their welfare."

Author's note: In September of 2016, the National Institutes of Health (NIH) held a workshop to discuss the use of nonhuman primates—in particular, monkeys—in biomedical research. Animal advocates, whose outcry over experiments that separated infant monkeys from their mothers had led Congress to call for the NIH's meeting, hoped for a frank and wide-ranging discussion, much like the landmark Institute of Medicine deliberations around chimpanzees.

Whether that happened is contested. After the workshop, many animal advocates accused the NIH of excluding critical voices and keeping conversation about ethics to a minimum. The NIH objected, describing the discussions as robust. To this reporter, both sides had a point.

Nobody from the animal advocacy community was included in the workshop. It was a glaring and unfortunate omission; they would likely have highlighted issues—shortcomings in certain primate models of disease, the performance of institutional committees tasked with protecting research animals—glossed over during the day's proceedings. That said, ethics were certainly not ignored; researchers spoke convincingly of improved welfare standards and the importance of using primates only when necessary.

An unresolved tension, though, exists between what's necessary and what's worth doing. And there was little acknowledgement of complex monkey cognition, further scientific evidence of which seems to emerge weekly: monkeys who are deeply self-aware, who communicate with something like language, who grieve for their dead. Why should a monkey be less morally significant than a chimp? That's the fundamental question—and as of now, the medical research community has yet to confront it.

I, COCKROACH

Between the ages of six and eight or so, when I was old enough to run around outside but too young to have cooler things to do, I spent quite a bit of time with insects. Not that I was especially into entomology or even science. Bugs were just something fun and animate to play with: I kept caterpillars, feeding them fresh leaves and cleaning their jars every few days, nourishing them to moth-hood. At the same time, with no sense of contradiction, I spent entire summer vacation mornings killing ants, spraying them with window cleaner, setting them on fire, or coaxing them to fight in bottles.

If that sounds sadistic, let me say that it wasn't done with a cruel spirit, or any memorable pleasure at the ants' discomfort. It was just something to do, and I don't think my experience was especially unusual, at least not among boys of my generation. Quite a few guys I've known can relate similar stories. Magnifying glasses are a fairly universal feature.

These days I don't much like to think of those ant-massacring mornings, but I did after reading about Backyard Brains, a Kickstarter-funded neuroscience education company. The company's flagship product is RoboRoach, a $99.99 bundle of Bluetooth signal-processing microelectronics that's glued to the back of a living cockroach and wired into the stumps of its cut-off antennae. Cockroaches use their antennae to detect objects; they react to electrical pulses sent through these nerves as though they have bumped into something, allowing children to remote-control them with smartphones. Other experiments involve measuring nerve activity in severed roach legs.

Given that few people spare a second thought to kitchen cockroach-stomping or classroom ant farms, the experiments might not seem too troubling. But using the insects like this, rather than killing them or watching them, is a different proposition. Some bioethicists have criticized Backyard Brains for encouraging children to think of living beings as tools, existing not for themselves but for our entertainment and edification. Those misgivings resonated with me. High school students might do this in biology classes—but should children on the low end of the company's suggested age appropriateness?

One of childhood's elemental lessons, learned in no small part through our immediate relationships with creatures less powerful than us, is how to think about and treat other living beings. There's no bright line, at least not then, between empathizing with animals and with people. I'm thankful that, in the end, my caterpillar-caring side prevailed over my ant-frying tendencies, and wonder if the instructive virtue of empathizing with insects might outweigh whatever educational gains can be had from steering them with an iPhone.

A note on the company's website does reassure customers that, though it's unknown if insects feel pain, anesthesia is used during procedures on cockroaches, and also on earthworms and grasshoppers involved in other experiments. This question of pain is an interesting one, and it opens up its own can of worms: As philosophers and scientists usually define the term, pain is intertwined with emotion, which in turn is intertwined with consciousness. You can't experience pain unless there's a you—a sense of self, an interior dialogue beyond the interplay of stimulus and involuntary response, elevating mechanics to consciousness.

Such sentience is quite unlikely in a bug, says Backyard Brains, and most people would likely agree. "It's very important to avoid anthropomorphising the cockroach with thoughts such as: 'If I do not want my own leg cut off, then the cockroach does not want its leg cut off,'" reads the site. And yet—do we really know this? A good scientist assumes nothing, and the possibility of insect sentience is rather more scientifically complicated than one might expect. In fact, there's good reason to think that cockroaches just might possess it.

Before dismissing bug consciousness out of hand—their brains are so tiny! And, they're bugs!—it's worth recalling that one of the first scientists to seriously consider the notion was Charles Darwin, who spent most of his adult life, even as he completed *The Descent of Man* (1871) and *On the Origin of Species* (1859), thinking about earthworms.

Earthworms aren't insects, of course, but for all practical purposes people usually lump them together in the realm of "invertebrate creepy crawlies with no meaningful inner life to speak of." Not so Darwin. His investigations of "how far [the worms] acted consciously," as described in his final book, *The Formation of Vegetable Mould through the Action of Worms, with Observations on Their Habits* (1881), run for more than thirty pages. In painstaking detail, he describes how earthworms plug the entrance to their burrows with precisely chosen and arranged leaf fragments, and how instinct alone doesn't plausibly explain that.

"One alternative alone is left, namely, that worms, although standing low in the scale of organisation, possess some degree of intelligence," wrote Darwin. "This will strike everyone as very improbable; but it may be doubted whether we know enough about the nervous system of the lower animals to justify our natural distrust of such a conclusion." Moreover, as the environmental philosopher Eileen Crist writes in her essay "The Inner Life of Earthworms" in the edited collection *The Cognitive Animal* (2002), Darwin doesn't simply describe the worms as unexpectedly sophisticated problem-solving machines. His descriptions implicitly acknowledge the realm of experience, of lives not lived by instinct alone, but with awareness and some ability to make decisions.

Some might accuse Darwin of anthropomorphizing, or of carelessly attributing human qualities to other animals. But assuming human uniqueness can be its own fallacy too. Darwin was, simply, scientifically open-minded. He granted that earthworms might be conscious of what mattered to them—objects in their immediate environment, the shape and temperature of their burrows—and made inferences on the basis of their behavior and what's known of intelligence in other animals, including humans. The worms couldn't speak, but the weight of evidence did.

That principle is fundamental to the consciousness many scientists now acknowledge in mammals and birds (and, depending on whom you talk to, in reptiles, amphibians, and fish). As for cockroaches, intimations of their experience could come from research on honeybee cognition, which has fascinated researchers ever since the zoologist Karl von Frisch's 1973 Nobel Prize-winning discovery of waggle dances, the complicated sequence of gestures by which honeybees convey the location and quality of food to hive mates.

Using methods designed to probe building-block fundaments of thought—Can a bee learn to follow green rather than red marks through a maze? How quickly does it apply that concept to geometrical marks rather than color? Do waggle dances convey only spatial information, or olfactory qualities too? How does a bee adapt when moved to an unfamiliar location or exposed to a new stimulus? Does that correlate with past experience?—scientists have assembled a portrait of extraordinary cognitive richness, so rich that honeybees now serve as model organisms for understanding the neurobiology of basic cognition.

Honeybees have a sense of time and of space; they have both short- and long-term memories. These memories combine sight and smell, and are available to bees independent of their immediate environments. In other words, they have internal representations of their worlds. They can learn to recognize patterns, and also concepts: above and below, same or different. They have simple emotions and beliefs, and apply those memories and concepts to their decisions. They likely recognize individuals. Those are qualities typically ascribed only to larger animals, with far larger brains, but life's challenges are universal: find food, don't become food, reproduce.

In a chapter of *Evolution of Nervous Systems in Invertebrates* (2007), the neuroscientists Randolf Menzel, Björn Brembs, and Martin Giurfa argue that, even if we've tended to assume that insects solve life's challenges mechanistically and without thought, there's now "considerable evidence against such an understanding."

Cognition is only one facet of mental activity, and not a stand-in for rich inner experience, but underlying honeybee cognition is a small but sophisticated brain with structures that effectively perform similar functions as the mammalian cortex and thalamus—systems considered fundamental to human consciousness.

The upshot of all this, thinks Menzel, is that bees themselves possess some form of consciousness. It's not a human consciousness, obviously, but certain features are likely to be common: a sense of awareness and intent, an "inner doing." And if it sounds like maybe someone's spent too much time with bees, it's also the belief of Christof Koch, the chief scientific officer at the Allen Institute for Brain Science in Seattle and one of the world's foremost investigators into the neural basis of consciousness. To Koch, consciousness is a function of neurological complexity, which bees and many other insects clearly have in abundance.

The nature of their consciousness is difficult to ascertain, but we can at least imagine that it feels like something to be a bee or a cockroach or a cricket. That something is intertwined with their life histories, modes of perception, and neurological organization. For insects, says Koch, this precludes the reflective aspects of self-awareness: they don't ponder. Rather, like a human climber scaling a cliff face, they're immersed in the moment, their inner voice silent yet not absent. Should that seem a rather impoverished sort of being, Koch says it's worth considering how many of our own experiences, from tying shoelaces to making love, are not self-conscious. He considers that faculty overrated. (For the record, Koch doesn't lose sleep over swatting a mosquito in the middle of the night, but neither will he kill insects when he can avoid it.)

It's impossible to know how cockroaches would perform in honeybee-style cognition tests, as few have even been attempted in other insects. Perhaps honeybees are exceptional. But perhaps not, says the ethologist Mathieu Lihoreau. His 2012 article for the journal *Insectes Sociaux*, "The Social Biology of Domiciliary Cockroaches: Colony Structure, Kin Recognition and Collective Decisions," co-authored with James Costa and Colette Rivault, is a must-read for anyone interested in these creatures.

Among the surprising—to me, anyway—facts detailed by Lihoreau, Costa, and Rivault about *Blattella germanica* (the German, or small, cockroach) and *Periplaneta Americana* (the American, or large, cockroach), found in kitchens and sewers worldwide, is their rich social lives: one can think of them as living in herds. Groups decide collectively on where to feed and shelter, and there's evidence of sophisticated communication via chemical signals rather than dances. When kept in isolation, individual roaches develop behavioral disorders; they possess rich spatial memories, which they use to navigate; and they might even recognize group members on an individual basis. Few researchers have studied their cognition, says Lihoreau, but cockroaches likely possess "comparable faculties of associative learning, memory and communication" to honeybees.

As to whether cockroaches possess a self, in the pages of *Cockroaches: Ecology, Behavior, and Natural History* (2007), cowritten by William J. Bell, Louis M. Roth, and Christine A. Nalepa, I happened on a reference to Archy, a popular early twentieth-century cartoon cockroach who said: "Expression is the need of my soul." Archy's inclusion was intended in fun, but there was a grain of truth. Cockroaches could very well possess a sense of self, and one that's perhaps not entirely alien to our own.

As for how they fare in Backyard Brains, I do feel fewer misgivings now than when I started learning about insect inner experience. After talking with one of the company's cofounders, a neuroscientist named Greg Gage, I came away impressed by his respect for the insects: if he doesn't grant them sentience, he at least recognizes the possibility that they feel pain and encourages customers to care for their animals when experimentation has concluded. I also suspect that a cockroach, by the standards of its own experience, can live as fulfilling a life in research as under a sink. And though I still don't like the thought of middle-schoolers doing the company's experiments, at least its sensibilities are rather more compassionate than that of most people, who regard cockroaches as fair game for squashing.

Gage also justifies the costs to their insects by the benefits of teaching children about neuroscience in experiments—and they have conducted research to show those benefits are tangible and real. It's a legitimate argument. At the very least, though, the costs should be calculated as accurately as possible, in light of the sentience that cockroaches—and earthworms, and crickets—very likely possess. Maybe that means using fewer of them, or housing them with an eye to the dark, warm, and moisture-rich conditions they naturally prefer. Perhaps it also means being thankful for them, and respectful, and actively trying to empathize with them.

I do wonder, however, whether both cockroaches and burgeoning scientific curiosities might be better served by studying cockroach behaviors and cognition in a less intrusive way. The Backyard Brains website admonishes against assuming a cockroach wants to keep its legs, but I don't think Darwin would agree. He'd see this as a question to be tested, and one can envision the experiment: If, after making a choice, such as going through one of two doors, a cockroach has a leg removed, will he or she be as likely make that choice again?

That's a bit gruesome, but other experiments in cockroach cognition would be fascinating. In the manner of honeybee learning experiments, squares and circles might be printed on the doors, with food placed behind one, in order to test whether cockroaches can learn shapes. To test spatial memory, the insects might be placed in mazes; and then, in light of research on how groups coordinate to find food, students might test whether large groups solve mazes faster. How do two roaches interact when they're familiar with each other? What about when they haven't met before? If they're closely related, or less so? Are these the results the same for males and females? And so on.

Let students do this sort of research, and leave the RoboRoach-ing and leg-cutting to those with an active interest in neuroscience, if only because cockroaches just might, however improbably, be happier that way. And to those who still consider that view soft-hearted and perhaps soft-headed, I offer this: according to Gage, remote-controlled roaches respond to commands only for a while. After that, they ignore the signals. As best as we can tell, they go where they want to go.

IV
ETHICS

In my professional life, I live in uneasy, oscillating relation to the world of ideas. My job is predicated on facts; spend enough time with them and grand narratives feel insubstantial and arbitrary. Why do men prefer women of a certain type? What's the secret of innovation? Another word for an empirically unverifiable explanation is myth.

In time, though, vision proscribed by the empirical feels myopic. After all, even a recitation of raw fact is shaped by values and principles; and certainly science's methodological limits are evident. It has enough trouble understanding what chickadees say, much less capturing the many essences of love. There's truth in poetry, too. And in the end, it's ideas that shape who we are, and thus the world—a truism become immediately literal in the Anthropocene, as people have named this modern age in which human activity profoundly influences Earth's biological and ecological processes.

It took a while to write that sentence: the Anthropocene is, after all, a contested fact. It's as much a political and ethical idea as a scientific one—sometimes the Anthropocene is defined as an age of human dominance, which is something else altogether, and a dangerous idea to accept. A good Anthropocene is one in which power is accompanied by humility and consideration, and our footprints guided by wildness and wilderness. More than any technological advance, these ideals will determine the future of life on Earth.

THE IMPROBABLE BEE

How does a bee find a flower? Perhaps, if he lives in a hive, another bee tells him where to go, but even that first bee needs to find the flower, and anyway most bees are solitary. So that bee flies off to forage, and if he's lucky it spots something of the right color, and that something turns out to be a flower. Not just any old flower, necessarily: some bees are generalists, but others specialized—for long-necked flowers, or maybe morning glories, or pea plants, or penstemons.

Will they always be successful? Of course not. But there's always a chance, and so bees fly. And what distances! Out in the tarmac sea of a stadium parking lot, the center of a Utah saltpan with no plants for miles on any side, on a third-floor balcony in an industrial Brooklyn neighborhood, where there's a box of linaria: a bee will come by, with eyes so powerful he can spot a fringe of red, a signal of readiness, at the throats of their fingertip-size blossoms.

Soon he's doused in yellow pollen, like a celebrant of Holi, the Hindu festival of colors, and carries it home. If these blooms are good, he knows, so nearby linaria might be ready too; he can smell these with antennae sensitive enough to detect part-per-trillion concentrations, to recognize single molecules floating in air.

What are the chances of that? Maybe some enterprising postdoc or apiarist has calculated them. Whatever they are, it boggles the mind; and yet it happens, again and again, an invisible equation of uncertainty, from which calculation—trillions upon trillions of times—the living world blossoms around us. You can reach out and touch it. Plant a flower and a bee will come.

I think of the technologies we would derive from bees—all those tech-press standbys of scientists who would use their eyes to make cameras, their antennae to detect bombs, their aeronautics to make, appropriately enough, new drones. Some researchers have even tried to make robot bees that could pollinate crops. I tend to be skeptical, though this could be useful enough. But will any of these bees ever make the flowers bloom?

THE ETHICS OF URBAN BEEKEEPING

Any day now, the apple trees on my deck will bloom, bringing with them the first honeybees of spring. It's a moment I'll greet with mixed feelings. To which bee-lovers everywhere may respond: How can anyone feel anything but good about honeybees? They're little gold-and-black life-bringers, booty-waggling symbols of industrious virtue, and now—after a decade of declines in commercial honeybee colonies—subjects of sympathy and concern. We all want to help the bees.

All of which is true, and honeybees are certainly not unwelcome guests. And yet: Most of them will be domesticated, belonging to colonies maintained by New York City's urban beekeepers. They're also competing for the same blossoms with wild bees. Indeed, urban honeybees have the potential to push out their wild brethren. "If you have lots of honeybees," says Dave Goulson, an insect ecologist at the University of Sussex, "it is bad for wild bees."

That bees other than honeybees and bumblebees even exist is a fact many people don't appreciate. I didn't until several years ago, when I stumbled across the extraordinary photography of Sam Droege. An entomologist with the United States Geological Survey, Droege literally revealed a new world to me: bees large and small, not just black and yellow but red and green and blue, an evolutionary jewel box of pollinators.

There are more than four thousand native bee species in the United States, many of them threatened and diminishing in number—a decline rather less remarked-on than the honeybee problems. Yet these wild bees also play a vital role in pollinating crops and the landscape writ large, in both the country and in cities. There are some two hundred wild bee species in New York City alone, and soon I started noticing them: leafcutters with pollen-doused foreheads, rice-grain-sized masked bees, and my favorite of all, the emerald-bodied virescent sweat bee.

Most of these wild bees are solitary, living alone or in small groups. They're not social like honeybees, who live in large colonies and, thanks to their marvelous communicative abilities, can descend en masse within minutes of a single bee finding a flower patch. Neither are they so aggressive as honeybees. Part of the reason native bees go unnoticed is that they're so unlikely to sting anyone (though sweat bees are indeed drawn to perspiration; after encountering Droege, I noticed that what I'd thought were flies chasing me on jogs were often bees).

These gentle loners, then, can easily be pushed to the side by honeybees, and much research describes how a large honeybee presence can exclude the natives. An unsettled question, though, is whether this is existentially bad or just inconvenient: Do native bees run out of food and decline in number, perhaps disappearing from the area altogether, or simply find another flower patch?

It's a difficult question to study in a methodologically precise manner, says Rachael Winfree, a pollinator ecologist at Rutgers University, but the crux is the availability of flowers. In suburban New Jersey, where Winfree works, there are plenty of blooms to go around. In other situations, though, such as urban areas with relatively few blooms early in spring or late in summer, floral bottlenecks may occur, with honeybees crowding out natives in a push for scarce resources.

Native populations can decline and, along with them, the pollination they provide to plants that go unvisited by honeybees or are not particularly well-pollinated by them. These include a cornucopia of crops, such as tomatoes and blueberries and squash, and a kaleidoscope of wildflowers and shrubs and trees. A locale without wild bees is a far less verdant place.

To be sure, this isn't a reason to condemn urban beekeeping, which is often pursued by people trying to enrich their pocket of the world rather than their own pockets, and provides a connection to nature that might not otherwise exist. And it's likely only a problem when a few dozen hives are concentrated in a relatively small area. A hobbyist's single backyard hive isn't as much of an issue.

Still, these conflicts do underscore that our honeybees don't exist in a vacuum. They're part of an ecology, sharing a neighborhood with other nectar-loving residents. To keep honeybee hives without paying any mind to bees already present, as with fashionable airport beehives that seem to be installed under the premise that untended green spaces are empty rather than teeming with life, runs contrary to the spirit of beekeeping.

Diane Thomson, a Claremont McKenna College ecologist who has studied interactions between honeybees and bumblebees, offers a compromise: Treat urban honeybees as a window into a wider world. In caring for and about them, care for wild bees, too. Plant pollinator-friendly native flowers, protect their habitat from overzealous landscapers, build places for them to nest, and tell other people about them. After all, the greatest native bee threat isn't honeybees, but people who don't think about nature.

THE WILD, SECRET LIFE OF NEW YORK CITY

Early this summer I met a friend for breakfast at a restaurant in Williamsburg, Brooklyn. While waiting for him to arrive I spent some time staring at a lot next door—a vacant lot, as the spaces are called, but also the block's one concentrated patch of greenery. It was scraggly and unremarkable, but a welcome respite from the neighborhood's densely packed brownstones and sunbaked pavement.

Afterward we walked to the old Domino Sugar factory, located on the banks of the East River, a once-industrial zone that's passed through gentrification and into the luxury development phase. Near the factory was a small park, previously the site of another vacant lot. On it the local hipsters had erected a giant teepee, outside of which a young woman explained that the park would soon be replaced by condos. When the bulldozers came, she hoped the teepee could be raised elsewhere. There was a vacant lot nearby, on the corner of Bedford Avenue and South 4th Street. "Right now," she said, "there's nothing there at all."

As it happened, that was the lot I'd just been enjoying. And though there were no buildings on the lot, and plenty of space for a teepee, there was certainly not nothing there. Compared with its surroundings, it was positively bursting with life: a pocket grassland where the hand of human development had skipped a beat. It was even a bit wild, a space where life exists independently and spontaneously rather than being paved under or converted to some approved purpose.

The young woman could hardly be blamed for not noticing it. Not many people do. We are in the habit of seeing untended nature as a sort of blankness, awaiting human work to fill it. It's right there in the name: vacant lot. A place where spontaneous life is invisible, or at best considered so many weeds, the term used to lump together and dismiss what thrives in spite of our preferences.

It's natural for us to elide the existence of what we don't notice, but when we do, we cultivate our own subtle form of emptiness. In cities, so-called vacant lands account for a sizable portion of our urban space: roughly 15 percent in most cities and about 6 percent in New York City. That's a whole lot of life we're not noticing.

Cities contain green spaces, of course. And New York contains two of the best: Central Park in Manhattan and Prospect Park in Brooklyn. But parks are destinations, manufactured for experience. They are places to go on a weekend, or a lucky work-free afternoon. Vacant lots are part of our daily surroundings. They are found in places where a property owner waits for the right deal, at the edges of roads and train tracks and parking lots, the liminal zones nobody can be bothered to beautify.

Even nature-lovers don't much care for these places. If they notice them at all, they see them as something to restore and reclaim. They're not a significant piece of nature, like a wetland or a forest. Instead they represent unfulfilled potential, as recreational spaces or ecosystem service-providers or simply a nature more to our tastes. Something ought to be done with them.

Yet there are other ways of perceiving vacant lot life. They have their own ecologies, well suited to urban conditions, and possess a surprising, even lovely, richness. They're where you can hear crickets and songbirds on the way to work. They're alive, and alive with the lessons of wildness: that humans are members of life's communities, a neighbor rather than an owner. "This ethical idea," wrote the environmental historian Roderick Nash, "may be the starting point for saving this planet."

It's a starting point that begins in our own neighborhoods. Environmentalists proclaim the virtues of living in cities, leaving space for nature outside them. That nature will be allowed to flourish, however, only if we respect it. It's a lesson learned not just from wilderness vacations, wildlife documentaries, or weekend visits to the park. It's a lesson to learn from everyday habitats, even in New York City. It begins with taking a close look.

When I first saw the Williamsburg lot, I recognized a few of the species—horseweed, for instance—from an ill-fated attempt at seed-bombing my own block's vacant lot. (The seeds failed to penetrate the weeds, the dense existence of which I'd somehow overlooked.) So my recognition of the Williamsburg lot's vitality was a coarse sort of appreciation: a step up from considering the lot empty, but far from acknowledging it in any detail.

"You can't understand anything about plants until you know what their names are," said Peter Del Tredici, a botanist at Harvard University's Arnold Arboretum and author of *Wild Urban Plants of the Northeast: A Field Guide*, indispensable to any city naturalist. "Even biologists don't bother to learn the names. They just consider them weeds. That lets you ignore and disregard them." I called the botanist Marielle Anzelone, founder of New York City Wildflower Week, an annual event that teaches people about the city's unappreciated flora, and especially its rare flowers. Anzelone is herself quite rare: someone who can walk down a street and put a name to just about every tree and flower and blade of grass.

We started our walk beside the fence along Bedford Avenue. Through Anzelone's eyes the undifferentiated green came into focus. Old stems of goldenrod ran along the fence. There was horseweed and low-growing field pepper and broadleaf plantain, anthered stems growing from ground-level leaf rosettes. Wood sorrel, pokeweed, boneset, and white snakeroot, its leaves whorled with trails left by larval insects. Japanese knotweed and Japanese brome, daisy fleabane, calico aster. Trailing over a cinder block were the vines of a Virginia creeper. Inside the lot, Anzelone found Boston ivy growing along the building, underneath the graffiti, along with sapling zelcova and pin oak trees. She identified more field plants: yellow-flowered black medic, a thorny clump of bull thistle, a patch of clover.

Whenever she found a native plant, one that could trace its lineage deep into the region's evolutionary history, a note of excitement crept into Anzelone's voice. For her, nativity is bound up with a sense of place, of rooting oneself beneath the froth of the contemporary. A clump of path rushes—a species that grows from sidewalk cracks—might conceivably trace its ancestry to seeds that took root when the great Laurentide ice sheet receded twenty thousand years ago, leaving behind the essential features of New York City's present geography.

Only a few of the plants in our lot were properly native, though. Most originated in Europe and Asia, arriving in the last few centuries with settlers, or imported by gardeners. Yet they have their own sort of nativity. They belong to a community of plants lately recognized by ecologists as characteristic of urban areas. They are, in the argot, ruderals. Disturbance specialists, wasteland-growers, found in abandoned spaces across New York City and much of the eastern United States.

These plants "live fast and die young," said Steven Handel, an ecologist at Rutgers University. They are annuals and short-lived perennials, able to flower and fruit and make seeds quickly, before someone cuts them or builds over them. Most have seeds lofted and dispersed by the wind. "Think of Brooklyn as an ocean of asphalt, with little archipelago islands of vegetation," Handel said. Some are dropped by birds, or carried in the fur of small mammals—mice and rats and feral cats—or on the shoes of local primates.

Moreover, these plants thrive in urban conditions, which are extreme. Like most large cities, New York City is a heat island, asphalt- and rooftop-trapped heat raising summer temperatures by an average of 5 degrees Fahrenheit. Vacant-lot soil can be nutrient-poor and root-stuntingly compacted; at some point in the recent past, a building stood where Anzelone and I do. It's also dry, receiving relatively little precipitation from rainfall that evaporates or flows into sewers rather than collecting as groundwater.

Out of this, ruderals spring. With each passing year, their growth and eventual decomposition enrich and soften the earth, fixing nutrients and enlivening the soil. In the process they sequester carbon and clean the air. They cool surrounding air, too, and help capture rainfall that otherwise swamps New York City's overloaded sewers. Those are ecosystem services, performed efficiently and free of charge, though that's just one perspective. Vacant lots serve others, too, creating food and habitat for the few creatures that live in the urban matrix, outside the protection of parks and conservation lands.

Anzelone and I sat for a while by a bumblebee-frequented patch of white clover. Something that's little appreciated, she said, is how vacant lots provide pollinators with food throughout the growing season: dandelions in spring, followed by clovers, then asters and goldenrod that bloom until first frost. Urban gardeners rely on pollinators, but it's not gardens that sustain them. It's places like this. Handel estimates that, in a decent-sized New York City lot, some five hundred insect species can be found.

I caught the electric green flash of a virescent sweat bee. Under a plantain leaf was a sow bug. Anzelone noticed a syrphid fly, part of an order that looks like bees. A black-and-yellow swallowtail butterfly flew overhead. House sparrows landed on power lines and foraged in a fallen *Ailanthus altissima*, or Tree of Heaven. In the suburbs, house sparrows are considered a songbird-threatening nuisance. To me they're musical neighbors. In my neighborhood I've also watched mockingbirds challenge crows, listened to cardinals sing their territorial bounds, thrilled to the blue-streaked flash of a kestrel falcon. They hunt the sparrows.

The vacant lot struck me as something like ecological graffiti. Not a cause of degradation, but a response to it. Mostly homogeneous in style, yet still a vibrant affirmation of life, and in some places beautiful, even special.

Across the river in Manhattan, where the pace and density of development rarely allows empty places to flourish, there are few vacant lots. Until recently, though, there was one immense and particularly notable vacancy: a mile-long, thirty-foot-wide stretch of elevated railroad track called the High Line, built in the 1930s, to haul freight down Manhattan's Lower West Side.

Soaring above meatpacking plants and warehouses, the High Line ceased operation in 1980, its last train delivering three carloads of frozen turkeys. The entrances were fenced off and locked up. The track itself, with only its steel-beam underbelly visible from street level, was largely forgotten, going quietly feral as the neighborhoods gentrified.

Now the warehouses are galleries, boutiques, and upscale lofts, and the High Line is an oasis of elegantly modernist paths lined with lush plantings conceived by the world-renowned garden designer Piet Oudolf. There are some 210 species of flowers and shrubs and grasses, the names alone a litany of vegetative delight: rosemary willow and northern blazing star, twilight aster and white turtlehead.

This new High Line is widely celebrated as a visionary example of green urban design, the triumphant reclamation of disused space. Not much of what was reclaimed now is recalled; mostly it's remembered as an eyesore. There's but one formal record of what lived in the decades between its uses as railway and park, a study published in 2004 in the *Journal of the Torrey Botanical Society* by a St. John's University biologist named Richard Stalter.

"The elevated railroad is a relatively inaccessible 'island' and my access to the High Line was through an artists' loft to the roof of an adjoining building. There, a ladder and rope provided a 'bridge,'" wrote Stalter. When he arrived, Stalter didn't see a wasteland. Through his eyes, the tires and bottles and trash helped create "a multiplicity of habitats" in what was an example of ecological processes also found along roadsides and, more to the point, bare rock and newly formed islands.

On the High Line, as on volcanic promontories rising from the sea, wrote Stalter, lichens and moss had taken a tenacious hold on stone and steel, growing and dying and thereby creating a first skein of organic matter. Eventually enough gathered for a few wind-blown grass seeds to take root, adding their own trace deposits to the nascent soil. Abetted by wind-blown dust, it soon could support even larger plants. By the time Stalter arrived, a layer of soil covered the High Line in depths ranging to nearly three feet.

And what grew in that soil! Stalter documented no fewer than 161 species of lichens and plants, split between two zones: a small area of shrubs and low-growing trees, and a larger area of grasses and flowers. These included many species found in the Bedford Avenue vacant lot—but unlike that and most other lots, the High Line was never mowed to keep vegetation in check, and perhaps for that reason accumulated many more.

Among these were nine species of aster and goldenrod, and also lesser-known plants like joe-pye weed. Art Shapiro, a former Staten Island resident who's now an entomologist at the University of California, Davis, described it as "a short, shining moment" when the High Line was "an absolute butterfly heaven." Stalter didn't inventory the invertebrates, but he did compare the floral diversity to other sites around New York City. Altogether, he wrote, it appeared to possess "one of the highest levels of species richness of any temperate zone urban environment in the region."

If the old High Line was a disused eyesore, it was also a botanical cornucopia, and all the more remarkable because it was untended. Nobody watered it, or added soil or fertilizer. Nobody weeded it. What's there now is a fundamentally different type of nature than before: a garden, and in a sense far less sustainable, requiring careful tending and more water than is naturally available.

It's also brought pleasure to millions of visitors, helped popularize a vibrant, less-manicured gardening aesthetic, and is far more pleasant than whatever would have been built had the High Line, spanning what's now some of the world's most expensive real estate, been torn down. There is joe-pye weed, and plenty of bees and butterflies, and birds, too. A few years ago, visitors watched peregrine falcons swoop down from their nest on the local Drug Enforcement Administration offices, which abut the High Line on 17th Street.

Yet when I walk the paths, something is missing. It's aesthetically pleasing but not inspiring. Nearly all of what lives there is what someone has chosen. Nature's spontaneity, testament to life outside human control, is largely gone. Of course people will argue that humans are part of nature, too, which is true, but there are different ways of being in nature. On the High Line, it's too easy to forget that verdancy is not the result of careful management, but of life's inexorable course, present wherever we don't suffocate it.

"On a planet increasingly permeated with human intentionality, areas we allow to be there for themselves, that we allow to become what they will, can stand in contrast to human hubris," wrote Roger Kaye in an essay for the National Parks Service. "They can counter the dominating presumption that everything exists in relation to us."

Kaye, a wilderness specialist and pilot for the United States Fish and Wildlife Service in Alaska, likely had a very different setting in mind. The principle still applies, though, in the city. Marvelous as the High Line is, it exists in relation to us. It's no longer wild.

About 10 miles southeast of Manhattan, as the pigeon flies, a ribbon of defunct railway echoes the High Line before the landscape architects moved in. It's part of the Rockaway Rail Line, running 3.5 miles through Queens and Brooklyn. The last train rolled in 1962; the rails were fenced off from the public, and nature left to go feral. Now it's described as a blighted eyesore, a neglected ruin. It's also a place to know better.

On a hot June day, unable to find a way up that didn't involve climbing a utility pole, I walked the elevated southern section at street level. Over its concrete edges poked red cedar and black cherry trees, multiflora rose, the brown seedheads of pigweed. Seeing me take photographs, a man in a sleeveless T-shirt with a faded cross tattooed on his shoulder introduced himself. "It's like a wildlife preserve up there!" he said, home to raccoons and possums and a great many birds. The latter were his favorites. Early in the morning, when he left for work at a plumbing supply store, cardinals and lately bluebirds were out singing.

That elevated section, said Handel, the Rutgers ecologist, resembles the old High Line: life filling and slowly softening an expanse of rock and metal. The northern section is also like the High Line in its feral nature, though topographically quite different. With richer conditions to start from, it's now a forest, old enough for trees to wrap trunks around the rails. Undergrowth is dense, in many places nearly impenetrable, but in a few spots at least it's possible to access on foot.

On another day, I visited with Anzelone, the New York City botanist. She characterized the denseness as a mix of woodland natives and nonnatives, many considered invasive. There were black cherry and white mulberry trees, horse chestnuts and pin oak and sassafras. Coiling some of the trunks and canopy were oriental bittersweet vines. There was Virginia creeper, and beds of poison ivy: a nuisance to humans, but manna for animals, who feed through winter on its fat-rich berries. Filling the understory were raspberry and porcelain berry bushes, nightshade and buckthorn—more manna.

Swallowtails flittered through openings in the canopy. A catbird thrashed in the brush, and a female cardinal perched nearby as we came upon several small trees in white plastic buckets. Tags identified them as part of the Million Trees Initiative, New York City's effort to expand urban tree cover. It took a moment for the absurdity to sink in, the implicit invisibility of the lushness around us. Someone was planting trees in a forest.

A local community group, Friends of the Queensway, hopes to turn the Rockaway Rail Line into a park called the Queensway. It would feature bicycle paths and gardens and meeting spaces; it would, argue the Friends, create jobs, improve public health, and simply be a fun place to go. Artistic renderings of the park look lovely. I'd be happy to have it in my neighborhood.

At the same time, I feel regret for what will happen should the railway become a park. Most of the verdancy will go. Trees will be beaten back, the poison ivy and bittersweet vines torn out, soil and water imported. Nobody, I suspect, will spare much thought for the birds.

Many scientists who consider these places from an ecological perspective see them as scrap heaps, dominated by invasive species that produce diminished, homogeneous ecosystems. The plants and animals living there don't need to be counted, and thinking otherwise risks a certain ecological Stockholm syndrome. After all, one could also celebrate the ruderal possibilities of a mine-tailing pond, or the parking lot borders of proposed developments in lush, ecologically rare Staten Island wetland forests.

Yet the Rockaway Rail Line, like the city's vacant lots, is to my eyes a treasure. Shapiro, the UC Davis entomologist, told me he saw vacant lots as "founts of evolutionary and ecological creativity" for dense urbanity, a space where megacity nature adjusts to our industrial presence, evolving the future's resilient ecosystems. To me they're simply rich, a place where nature's chorus refreshes the soul and you can feel a vital pulse of wildness. They're already homes, too. Life pushes back in vacant lots, asserting membership in a community of which, as Aldo Leopold put it, *Homo sapiens* is not a conqueror but "a plain member and citizen."

What it means to be a good neighbor to these communities, respectful and sharing, is a koan for the Anthropocene age. Ultimately it's our communities who decide what happens in their backyards. Confronted with a vacant lot, most people will prefer parks and gardens—somewhere to grow food and enjoy a neatly tended nature, one that's also not fenced off and gathering trash.

But these might be imagined from the perspective of vacant lot life, even inspired by it. As we think about developing our vacancies, we can get to know them: enjoy their plants and animals, challenge ourselves to plan in ways that support as much life as now exists. What we mow out of habit rather than need can be left to grow free. A few places might even be protected—pocket wilderness parks, with trails and benches, to go with our gardens. Maybe some of that Rockaway rail line poison ivy can be left for the birds.

Since visiting that Williamsburg lot with Anzelone, I've made my city sojourns with an eye to ruderals and the untended. There's far more life than I ever realized, and our own landscaping often suffers from the comparison. On one neglected side of a street, untended green will overflow, buzzing and blooming; on the other, a few desultory shrubs and factory-grown flowers stand in deserts of mulch and inch-high grass.

Recently I went back to Williamsburg, curious to see the lot in late summer. It's now a hole in the ground, with construction starting on an apartment building. It was difficult to begrudge. People need homes, and construction jobs. Later that day, I stopped by a vacant lot in my neighborhood. Until recently, it was a glorious profusion of Queen Anne's lace and chicory and milkweed, food for monarch butterflies that pass through New York City on their migration to Mexico.

Now it's mowed flat. There didn't seem to be any construction. Maybe someone complained to the Department of Sanitation, the city's vacant lot-mowers, about the unkempt growth. It's not unkempt now. It's obliterated. Shorn, silent, and motionless, it's something close to nothing. Yet amid the stubble were stalks of fallen milkweed. Their pods were bursting with seed.

EARTH IS NOT A GARDEN

Several years ago, I asked a biologist friend what she thought of a recently fashionable notion in environmentalist circles: that pristine nature was an illusion, and our beloved wilderness an outdated construct that didn't actually exist. She'd just finished her shift at the local boardwalk, a volunteer-tended path through a lovely little peat bog that formed after the last ice age, near what is today eastern Maine's largest commercial shopping area.

After a moment's reflection, she said this was probably true, in an academic sense, but she didn't pay it much mind. The fact remained that places such as the bog, affected by human activity, were special, and ought to be protected; other places were affected far less, but they were special and needed protection, too.

It was a simple, practical answer, from someone who'd devoted much of her life to tending the natural world. I find myself recalling it now that the ideals of conservation are under attack by the movement's own self-appointed vanguard: the ecomodernists, a group of influential thinkers who argue that we should embrace our planetary lordship and reconceive Earth as a giant garden.

Get over your attachment to wilderness, they say. There's no such thing, and thinking otherwise is downright counterproductive. As for wildness, some might exist in the margins of our gardens—designed and managed to serve human wants—but it's not especially important. And if you appreciate wild animals and plants for their own sake? Well, get over that, too. Those sentiments are as outdated as a daguerreotype of Henry David Thoreau's beard, dead as a dodo in an Anthropocene age characterized by humanity's literally awesome domination of Earth.

That humanity has vast power is true. Human purposes divert roughly one-fourth of all terrestrial photosynthetic activity and half its available fresh water. We're altering ocean currents and atmospheric patterns, and moving as much rock as the process of erosion. The sheer biomass of humanity and our domesticated animals dwarfs that of other land mammals; our plastic permeates the oceans. We're driving other creatures extinct at rates last seen 65 million years ago, when an asteroid struck Earth and ended the age of dinosaurs.

By midcentury, there could be 10 billion humans, all demanding and deserving a quality of life presently experienced by only a few. It will be an extraordinary, planet-defining challenge. Meeting it will require, as ecomodernists correctly observe, new ideas and tools. It also demands a deep, abiding respect for nonhuman life, no less negligible than the respect we extend to one another. Power is not the same thing as supremacy.

If humanity is to be more than a biological asteroid, nature lovers should not "jettison their idealised notions of nature, parks and wilderness" and quit "pursuing the protection of biodiversity for biodiversity's sake," as urged in a seminal essay coauthored by Peter Kareiva, chief scientist at the Nature Conservancy, the world's largest conservation organization. Nor can we replace these ideals with what the science writer Emma Marris imagines as "a global, half-wild rambunctious garden, tended by us."

Well-intentioned as these visions might be, they're inadequate for the Anthropocene. We need to embrace more wilderness, not less. And though framing humanity's role as global gardening sounds harmless, even pleasant, the idea contains a seed of industrial society's fundamental flaw: an ethical vision in which only human interests matter. It's a blueprint not for a garden, but for a landscaped graveyard.

Ecomodernism is not precisely new. Rather, it crystallizes arguments that have percolated through conservation for the past couple of decades and hit the zeitgeist after the publication of Marris's *Rambunctious Garden: Saving Nature in a Post-Wild World* and *Love Your Monsters: Postenvironmentalism and the Anthropocene*, a collection of essays produced by the Breakthrough Institute, a green modernist-leaning think tank based in California. Add Kareiva's prominence at the Nature Conservancy, the media hook of a paradigm shift, and a general frustration—shared by ecomodernists and their critics—with conservation's inability to halt ecological destruction, and the stage was set for what the US journalist Keith Kloor called a battle for the soul of environmentalism.

To this battle, the ecomodernists bore strategic and tactical arms. In their eyes, conservationists were too gloomy, fixating on stories of loss: polar bears on melting icebergs, ecosystems forever changed. Nature is resilient, they argued. Forests grow back. Polar bears might handle the heat after all. Also, conservationists should be more realistic, acknowledging the impracticality of their hopes, and more humanistic. Too often had nature's protections come at human expense.

"Conservationists will have to embrace human development and the 'exploitation of nature' for human uses," wrote Kareiva and his coauthors, Michelle Marvier, an environmental studies professor, and Robert Lalasz, the Nature Conservancy's science communications director. Ecomodernists called for corporate partnerships and an emphasis on ecosystem services, by which nature is measured and commodified according to the benefits it provides us. They also chided conservationists as technophobes, slow to celebrate technologies that promote prosperity while shrinking our ecological footprints.

In themselves, many of these ideas were not so radical. They represented a twenty-first-century update of a managerial approach already evident in everything from fishing limits to environmental impact statements. What made ecomodernism so controversial was the implicit ideological shift: that human interests should be elevated above all. The justification for this amounted to a rewrite of ecological history.

According to ecomodernists, mainstream conservationists are hopelessly fixated on a nineteenth-century vision of virgin wilderness unsullied by human hands. Consumed by nostalgia, they've failed to grasp the historical extent of humanity's ecological influence, practicing a wilderness worship that's not only ineffective but also delusional. "Conservation cannot promise a return to pristine, prehuman landscapes," Kareiva and colleagues wrote. "The wilderness so beloved by conservationists—places 'untrammelled by man'—never existed, at least not in the last thousand years, and arguably even longer."

Peering through sepia-tinted glasses, conservationists supposedly ignore less-pristine sorts of nature, the landscapes and habitats intertwined with our activities. Recalling a childhood spent playing in second-growth forests of the Pacific Northwest, Marris wrote that "ecologists and conservationists have long ignored such forests. In thrall to the lure of 'pristine wilderness,' they shunned second growth."

It's an odd statement, given that a Google Scholar search of "second-growth forest" returns more than fifty-four thousand hits; and it's easy to think of several conservationists whose environmental consciousness blossomed in such "ignored" spaces. The poet-environmentalist Gary Snyder grew up in Marris's own neck of the woods, on a "stump farm"—a term derived from harvesting stumps left behind when giant first-growth trees were felled—north of Seattle. Bob Marshall, the forester and activist whose efforts led to the Wilderness Act of 1964 and America's federally designated wilderness system, cautioned in 1930 that "unreachably high" standards of purity would leave second-growth forest unprotected. And Aldo Leopold, a father of modern environmentalism, whose seminal book *A Sand County Almanac* chronicled nature on a Dust Bowl-era prairie farm, is an especially interesting figure.

A hunter and forester, Leopold was steeped in the managerial tradition of nature as resource, of trees and fish and pheasants existing for our use, albeit responsibly. Yet he came to consider this ideal insufficient, a transformation captured in "Thinking Like a Mountain," his essay on killing a wolf. "After seeing the green fire die" in the wolf's eyes, he wrote, "I sensed that neither the wolf nor the mountain agreed with such a view."

Then there's Thoreau himself, conservation's patron saint and a favored target for Anthropocene ideologues, who mock him for extolling the virtues of wildness while living within walking distance of town. If anything, this should make him even more relevant. Thoreau found "a civilisation other than our own," and proposed that "in wildness is the salvation of the world," in second-growth forest and alongside railroad tracks. But this is inconvenient for ecomodernism, just as it's inconvenient to note the actual definition of the "untrammelled" condition treasured by conservationists.

It does not mean untouched. It means unrestricted. Wilderness, as formally enshrined in the Wilderness Act, is simply a place where nature's processes have not been severely diminished by human activity. It's best understood as a scale of wildness, a term originating in the Norse word for will. Wildness is—per Thoreau, together with the environmental historian Roderick Nash and generations of conservationists—that which is self-willed. It is free and autonomous, existing independently of human control. That's what impassions so many conservationists: not just the Arctic National Wildlife Refuge or the Amazon's jungle heart, but also the wild life of their own everyday landscapes, their own bog boardwalks. A few might fixate on the pristine, but most possess a love of wildness and a far more pragmatic appreciation of wilderness.

Never mind that; the caricature of hair-shirt conservationism is a necessary foil for ecomodernism's second premise: that twenty-first-century human activity merely continues what we've been doing for millennia. Those ostensibly pristine pre-industrial landscapes that hold conservation in their thrall were already transformed, in light of which the wilderness ideal is nonsensical.

"Some environmentalists see the Anthropocene as a disaster by definition, since they see all human changes as degradation of a pristine Eden," wrote Kareiva, Marris, the environmental scientist Erle Ellis, and the ecologist Joseph Mascaro in a 2011 *New York Times* editorial. "But in fact, humans have been changing ecosystems for millenniums." The Breakthrough Institute's founders Michael Shellenberger and Ted Nordhaus struck a similar note in *Love*

Your Monsters: "The difference between the new ecological crises and the ways in which humans and even pre-humans have shaped non-human nature for tens of thousands of years is one of scope and scale, not kind."

The Amazon is a favored example of this mass alteration of nature. The discovery of prehistoric planting beds and irrigation channels beneath presumably primal forest sparked a flush of excitement a decade ago. Those discoveries, best known from Charles Mann's *1491: New Revelations of the Americas before Columbus* and from work led by the anthropologist Michael Heckenberger, seemed to rewrite ecological history. Far from being untouched wilderness, the great Amazon rainforest, Earth's delicate lungs, had until recently been developed! "The whole notion of a 'virgin rainforest' may be erroneous," wrote Kareiva in 2007 in the journal *Science*.

But that picture is highly controversial. Although some parts of the Amazon were indeed populated, these were comparatively small: patches of intense modification strung through a vast wilderness. Certainly Amazonia was not domesticated at anything remotely like an Anthropocene scale. Some archaeologists have even suggested that the Amazon of the earthwork builders was a relatively dry, savannah-like place, making their accomplishments less dramatic. They practiced local farming, not landscape engineering.

These aren't just archaeological quibbles. Blurring the lines between limited human impacts in the past and extensive, Anthropocene-scale human activity today obscures the large and small wildernesses that still exist, from Southern Ocean deeps to boreal forests and much of the Amazon. Conflation also elides differences between lifeways. The limited, attuned management of nature—fires that are intentionally to mimic regional fire regimes, or agricultural systems dictated by existing hydrological cycles—is rated the same way as activities that overwhelm those ecologies.

Not that indigenous societies were always harmonious stewards with light footprints. There are many examples to the contrary, most notably the large-animal extinctions that followed once stone-age humans arrived in the Americas and Australia. Yet those extinctions hardly constitute the end of wilderness, or fall on the same spectrum as industrial-scale development. Once we outcompeted 20-foot-tall giant sloths and saber-toothed tigers; now we have trouble sharing even with kangaroo rats and tiger salamanders. That's the difference between the transformations wrought by several million people and by 7 billion, with drastically different resource requirements, and it's obscured by a narrative of human omnipresence.

This narrative is linked, says the environmental law professor David Johns in an essay in *Keeping the Wild: Against the Domestication of Earth*, "to the idea that significant human presence and impact means that humans are in charge." It's an exercise in modern mythmaking, the normalization and self-justification of a human-dominated, wilderness-free Anthropocene, in which those pesky conservationists need to get with the program. As Marris argues in *Rambunctious Garden*: "We are already running the whole Earth."

All this might strike some as a tempest in a rooibos teacup: an internecine spat. But ideas matter, especially when they involve such fundamental questions of how we see ourselves in relation to other life. Discarding wilderness and downplaying wildness leads to some troubling ethical places. It comes from a troubling place, too: the environmental philosopher Eileen Crist likens ecomodernism to the mentality that put humans atop the medieval Great Chain of Being, and kept them there until Charles Darwin.

The garden metaphor is particularly loaded. A garden is not an ethical place. Life and death occurs at a gardener's caprice. Plant or cut, tend or kill, include or exclude: it's an exercise of morally unconstrained will, fine in a backyard but requiring boundaries both literal and philosophical. "Once we shift to a gardener's mindset, it gives us too much freedom to do whatever we want," says the bioethicist Gregory Kaebnick. "And I'm speaking as an avid gardener."

In a garden, there isn't necessarily a sense that life has any value apart from what we assign. Neither individual beings nor larger entities—populations, communities, species, ecological processes—are intrinsically worthy of respect. The gardener ethic can't account for that. What Leopold so eloquently advised, that we think of ourselves not as conquerors of life's communities but as "plain members and citizens," goes out the window.

Moreover, the values we do assign are inevitably weighted toward what we've planted and controlled. "Civilisation . . . is the garden where relations grow," wrote the poet Howard Nemerov, a telling phrase quoted by Gary Snyder in *A Place in Space*. "Outside the garden is the wild abyss." This habit of mind is already deeply embedded in our development-oriented society. What's unmanaged is devalued, if not downright invisible. Nobody walks straight through a flower bed, yet we trample a vacant lot's so-called weeds without even noticing them. In a gardened Anthropocene, nature is at risk of becoming an abstracted aesthetic, interesting largely because we can find our own reflection in it.

Hence restoration to pre-industrial ecological baselines is considered impractical, but so-called Pleistocene rewilding—parks managed to contain analogues of million-year-old ecosystems—is celebrated. Trying to keep species from going extinct is old hat, but "de-extinction" to biotechnologically re-create them is fashionable. Such projects are worthwhile, but they reflect a self-centered sense of Anthropocene nature that easily turns toxic.

An egregious example comes from a National Public Radio blog post from this January titled "A Human-Driven Mass Extinction: Good or Bad?" The writer, Adam Frank, riffed on a *New Scientist* interview with the ecologist Chris Thomas, who observed that some lineages are adapting to the human activities now driving so many others extinct. "How does that make you feel? How should that make you feel?" asked Frank about the imminence of mass extinction. "The answer to this question depends mightily on what you think of as nature and where you think we fit into it."

That line of thought would never be applied to human affairs—"A city was bombed last night, but other people are building shacks with the rubble! Good or Bad?" The very notion is risible. Human lives, after all, have intrinsic value. But by the logic of ecomodernism, nonhuman lives do not. Mostly they represent preferences, or services. It's the exploitative essence of colonialism applied to nature, an enshrinement of the very attitudes that turned humanity into a biological asteroid. To quit protecting nature for its own sake, to judge conservation by the extent to which it furthers human interests, is to reconfigure our relationship to Earth as that of an empire to a resource colony. Earth: it's all about us.

Wilderness and wildness are the opposite of that. As ideals, they embody respect for nonhuman lives, recognizing that they don't exist solely in relation to us. Entering a wilderness, wrote Nash, we realize that we're not in a playground, but someone else's home. It opens us to perceiving the intrinsic value of other lives and the importance of sharing.

"Conservation is driven by asking: 'Who are the members of my community?'" says Kieran Suckling, executive director of the Center for Biological Diversity in Arizona. "It is an act of humility. It requires you saying, 'What's important is not only my desires and needs. It's also what other creatures want and need.'"

Which isn't to say that humans can or should cause no harm at all, much less cease activity. It's impossible not to have a footprint. But we can think about where we put our feet, cultivate a sense of modesty rather than lordship, and let courtesy and respect guide us. It's a habit of mind useful in affairs with our own species, too. "The lessons we learn from the wild," writes Snyder, "become the etiquette of freedom."

Easier said than done, of course. Ecomodernists would retort that it's unrealistic: those lovely ideals might have won a few battles, but conservationists are losing the war. New ideas are needed. Leaving aside the dubious proposition that principles should be abandoned when they're difficult to practice, quite a few of their criticisms ring uncomfortably true.

Clearly something more is needed. Were conservation's laurels enough to support our weight, Earth wouldn't be headed for a mass extinction. And it's true that conservationists have too frequently relied on shopworn platitudes. Everything is very manifestly not connected to everything else; the extinction of a butterfly won't bring down the whole ecological edifice. Likewise, prejudices against nonnative species can be counterproductive; if nonnative vegetation were eradicated, a great many butterflies would starve. The language of fragility and irreversibility grows tiring, too. Life can be fragile, but it can also be remarkably resilient.

The green modernist emphasis on ecologically sound development is also welcome, even if its boosters strike some creepy notes. "Putting faith in modernisation will require a new secular theology," write Shellenberger and Nordhaus, and "require a world-view that sees technology as humane and sacred." The emerging field of conservation finance, a nature-focused spinoff of impact investments geared for social benefit, has tremendous potential. Billions of people want more prosperous, secure lives, and how they achieve them will determine nature's baseline conditions across much of our planet.

However, without ideals of wilderness and wildness as guides, the compass spins astray. Pursuing intensification—the green modernist goal of more productive agriculture and denser cities, presumably leaving more space for nature—requires ethical direction. Otherwise the promise is merely of intensification across ever more of Earth's surface. Embracing technology is similarly open-ended counsel. There's a difference between crops engineered for high-yield drought resistance and those designed to withstand high doses of multiple pesticides.

Making such distinctions requires critical thinking about technology and development, not Kareiva and Marvier's request that conservationists quit "scolding capitalism," as if capitalism were some transcendent entity rather than a subject of constant debate. Equally unhelpful are the false dichotomies of Shellenberger and Nordhaus, who write that "living in a hotter world is a better problem than living in one without electricity."

Ecomodernists make a point of celebrating urban nature, which is good. Nonhuman city life should be appreciated as a matter of principle. Bob Marshall wished we would "learn how to treat even the smallest elements of the natural world with respect," even squirrels and scraggly inner city elms. But as a tool for preserving life's richness, urban nature is limited. Witness the global survey, published earlier this year in the *Proceedings of the Royal Society B*, of birds and plants living in cities. Urban locales contained, respectively,

20 percent of the birds and 5 percent of the plants that would be found in corresponding nonurban locales. Take away a city, then, and there's twenty times the number of plant species, and five times as many birds. Yet this somber statistic was greeted in some quarters as good news: cities can support life!

That Panglossian impulse characterizes many of the examples given by ecomodernists as evidence of resilience: rare salamanders that, having lost their natural habitat, become specialized to live in cattle tanks; rainforests regrowing on abandoned agricultural land and that contain half the species that previously lived there. Chernobyl is another, iconic example of ostensible Anthropocene resilience: even after a nuclear meltdown, life thrives! But Chernobyl's great and terrible lesson is that a nuclear meltdown is better for nonhuman life than having us around. The best garden is not a garden at all. Earth's plants and animals still need wilderness, big and small.

Some of that wilderness, of course, and much wildness, will need to exist in places where humans live too. It's great that the Nature Conservancy is working with the mining giant Rio Tinto to reduce its impacts in the Gobi desert, and giving advice on how to site dams in Columbian rivers to minimize their ecological harms. Damage control is vitally important. But it shouldn't be conservation's highest aspiration, and it's hardly a strategy on which to bet Earth's future.

To recommend, as Kareiva does, that we quit "pursuing the protection of biodiversity for biodiversity's sake"—or, in less sterile terms, life for life's sake—and measure conservation "in large part by its relevance to people" isn't a bold new vision. It's surrender. It's also impractical: When can a line ever be drawn? When can a bit more prairie not be paved, or a river be left undammed—since damming and paving creates novel ecosystems of sorts too, and testifies to nature's ability to adapt? Against this, ecomodernism would draw lines with ecosystem services: grasslands support pollinators, undammed rivers sustain commercial fish stocks. But can this strategy succeed on large scales? It might be possible to have rich services—clean air and water, productive working landscapes—with a minimum of biodiversity.

As the biologist John Vucetich points out, the modern world aptly demonstrates that we can have human prosperity in the absence of once-abundant life. "There are so few black-footed ferrets on the planet that we've already proved we don't need them," he says. "We'll get along fine without swift foxes, or wolverines. You can list hundreds of species we can get along fine without."

Vucetich sees the fight over ecomodernism as a clash between two visions of sustainability: one that exploits nature as much as we like without infringing on future exploitation, and one that exploits nature as little as necessary to lead a meaningful life. "It's hard to imagine those two worldviews would lead to the same place," he says. "I think they would lead to wildly different worlds."

Conservation's successes are not just national parks and protected areas but also the many places where human appetites are restrained so other life might flourish: forest pockets at the edge of town, wetlands spared conversion to shopping malls. Defending these is the bread-and-butter work of conservation, and this needs ideals.

"People volunteer to be on committees to study the mining proposals, critique the environmental impact reports, challenge the sloppy assumptions of the corporations, and stand up to certain county officials who would sell out the inhabitants," wrote Snyder of his neighbors' efforts to guard their second-generation Sierra Nevada foothill forests from irresponsible mining and logging. Motivating them in this unglamorous work is not self-interest, he writes, but "a true and selfless love of the land." It's hard to imagine ecomodernism being so inspirational, or for that matter effective. Despite caricatures of conservationists as unwavering zealots, in practice conservation is an exercise in negotiation and compromise—and no decent compromise is ever reached by meeting halfway at the outset.

Here I think of another Maine example: the ongoing restoration of the Penobscot River, the second-largest river in the northeastern United States, which for much of the past 9,500 years conducted vast runs of migratory fish between the Atlantic Ocean and the river's 7,700-square-mile watershed. For the past two centuries, all but a few miles were locked behind a series of dams. Those migrations, which the historian and biologist John Waldman says once made Eastern rivers "run silver," had been winnowed to a few pitiful remnant trickles.

Over the last decade, an alliance of conservation organizations and government agencies struck a deal with electricity companies: they'd purchase three dams, removing two and creating a passage around the other, while one dam would remain and increase its power generation. The agreement was precisely the sort of thing that ecomodernists champion: a balance between commercial interests and ecological values, providing energy while protecting life.

The Nature Conservancy helped to make the restoration possible, and rightly celebrate it on their website. But they weren't the only ones involved. Negotiations were realized only because conservationists and fishermen had previously fought for decades to prevent further dam construction. The commercial value of restored fisheries and recreational opportunities was part of the dam removal argument, and helped to sell the project at a policy level. But the animating force that sustained so many people through years of town meetings, impact statements, and potluck fund-raisers was a love for what the Penobscot once was.

One woman I know, an artist and wildlife ecologist who makes woodblock prints of river animals for local schoolchildren to use in art class, describes the project as a rebirth. Sometimes she wades the shallows, feeling in the mud for old stone sinkers left by Penobscot Indians in pre-industrial times. The Penobscot Indian Nation has supported the restoration too, treating it as an opportunity to heal the degradation of waters once central to tribal life. And though they have no illusions of the river returning to its former richness—after all, negotiation left one dam intact—they're inspired by the vision. They're also stirred by an intuitive sense of wrongness. For the Penobscot's life to have been so heedlessly truncated was just not right.

It's these feelings, wilderness-impassioned and modest and respectful, that ultimately made compromise possible. It wasn't ecological cost-benefit analyses, nor was it some sense of humanity as planetary gardeners. Take away wilderness, both its reality and its ideal, and the dams would still be choking the river.

ADD A FEW SPECIES. PULL DOWN THE FENCES. STEP BACK.

The early twenty-first century is a soul-searching moment for conservation. With each new report of vanishing species and dwindling biodiversity, the last century's great successes grow distant. Fundamental ideals and assumptions, in particular our cherished notions of wilderness, often feel ill-fitted to a crowded planet of more than 7 billion people.

There are more stories of loss than of hope—and a sense that something new is needed. Perhaps that something is *Feral: Rewilding the Land, the Sea, and Human Life*, George Monbiot's deep-throated paean to nature's joys, resilience, and great still-remaining possibilities.

Feral, released last year in the United Kingdom and now arriving on North American shores, arose from Monbiot's own soul-searching. A zoologist-turned-investigative-journalist, his Wikipedia biography includes beatings by military police, land-squatting showdowns in London, and a near-death brush with malaria contracted while living among East African nomads. Monbiot writes that, as his firebrand youth turned to middle age, with the confinements of home renovation and resort vacations, he experienced "ecological boredom."

Monbiot felt disconnected from nature. At the same time, he was unsure how to reengage. Environmentalism taught him what not to do—damage, pollute, waste—but offered few positive visions. "We know what we are against," writes Monbiot. "Now we must explain what we are for."

His words resonate with what's sometimes called "new conservation," best known from the writings of The Nature Conservancy head scientist Peter Kareiva and the author Emma Marris. They've called for a more human-friendly, forward-looking approach to conservation—one less fixated on often arbitrary and unattainable visions of pristine wilderness and more focused on supporting sustainable development and convincing big business of ecology's utilitarian values. Sometimes absent from the suggestions, though, is a certain vitality.

Ecosystem services and working landscapes are worthy goals, but where's the inspiration—or what Monbiot calls "that high, wild note of exultation"? For him, it's in wildness, a sense of communion with oystercatchers and boar, and the jostling of flowers. It's also in our ecological history—what and who lived where we are now until just a few thousand years ago.

Here, Monbiot veers from the new conservation in a subtle, fundamental way. In his own United Kingdom, what passes for pristine is, in fact, often blasted. Heath-covered Cambrian mountain slopes, widely considered the great wilderness of Wales, were flayed by centuries of sheep grazing. Their bare scrim of life is now perceived, somehow, as a bounty. Ditto the North Sea, where fish populations of a century ago, already ravaged by overfishing, are defined as a target for presumably recovered stocks.

If conservation elsewhere isn't always so myopic, shifted baselines are universal. "The state of nature is a state of almost inconceivable abundance," Monbiot writes. "Ours is a dwarf and remnant fauna, and as its size and abundance decline, so do our expectations, imperceptibly eroding to match the limitations of the present."

The book threatens to become yet another narrative of loss, but it never ventures far in that direction. Records of six-mile-long fish shoals and fossil footprints of vanished cranes are not laments. They are manna for imagining what life's resiliency still makes possible. Depleted regions may again teem with wildlife, and in many places—including rewilded patches of Cambria and Central Europe—they already are. "Rewilding offers the hope of a raucous summer," he exults.

An ever-evolving notion, "rewilding" comes in many forms. It includes releasing captive animals into the wild, as with beavers in Wales, where they once created landscapes speckled with ponds and wetlands. Or rewilding can be more ambitious, restoring suites of once-present plants and animals to ecosystems large and small. Oostvaardersplassen, the Netherlands' Pleistocene-styled park, is roughly the size of Manhattan. As a strategy, rewilding is formulated by contemporary conservation biology in terms of "cores, corridors, and carnivores."

Monbiot's rewilding encompasses these, but with differences. He emphasizes the richness of ecological interactions, the flow of energy and nutrients across time and form, and also—crucially—the essence of wildness itself. "Rewilding, to me, is about resisting the urge to control nature and allowing it to find its own way," Monbiot writes. "The ecosystems that result are best described not as wilderness, but self-willed."

Add a few species, pull down the fences, step back: this is not a fussy sort of conservation, then, bearing lists of favored species and blacklists of invasives, struggling to produce a historical snapshot. The past is inspiration, not blueprint. Neither is it a domineering conservation, managing nature to meet our demands. Services might result, but they're not the point. The ethos is not of human primacy, but a muscular, can-do humility. Things will fix themselves if we let them.

Between the ideal and the reality, however, falls the shadow. In *Feral*, Monbiot profiles two men: Ritchie Tassell, a forester rewilding the Cambrian heath, and the Welsh sheep farmer Dafydd Morris-Jones. Tassell's vision resonates with the author, yet Monbiot deeply admires and respects Morris-Jones, whose family has lived in the same Cambrian mountainside home since the nineteenth century. Morris-Jones weighs sheep by eye and loves the land; he also sees rewilding as a threat to his livelihood and community. There's no room in the forest for farmers.

The idea of pushing people like Morris-Jones aside is "intolerable," writes Monbiot, an avowed progressive and indigenous-rights activist. He wrestles with the tension. One way to ease it, he believes, is reforming British agricultural policies that are now unproductive, destructive, and expensive. They subsidize landowners who barely work their land but are required, as a condition of receiving subsidies, to prevent wild nature from returning to it. Fix those policies, and there's space for both farmer and rewilder. Of course, differences won't always be so easily reconciled, but the principle is universal. If rewilding is to happen, it must be the people's choice—an act of democracy, not conservation fiat.

For people to choose nature, though, they must also love it: not an ecosystem service or some abstraction of biodiversity (a word that, blissfully, appears rarely in *Feral*) but the living, breathing experience. Monbiot's own love, and his eloquence in expressing it, makes *Feral* not only interesting but a pleasure to read.

A sponge-covered king crab pried from the seabed "bulged with the suggestion of muscle like a Roman suit of armour." Chickadees sing in hills "patched with the small dark shadows of cloudlets," and oaks "put out embryo leaves as minutely serrated as mouse paws." Monbiot's prose incandesces with detail.

This is what conservation is for: beauty, richness, vibrancy. Life. We'll never recover all that was lost, but that's okay. What matters is the celebration and journey of revival. To rewild is to know where we are, learn its history, and ask: Could there be more? Can I help? These questions belong in backyards and roadsides, mountains and seas. It's no longer conservation, but revitalization.

FERAL CATS VS. CONSERVATION: A TRUCE

Feral cats are an invasive species that kills billions of wild animals each year, threatening entire species with extinction. They're also extremely difficult to control—and, well, they're just so damn cute. How could anyone eradicate as a pest a critter who just wants to play with packing peanuts and hide in boxes?

But the war between cat-lovers and bird-lovers may have found its compromise: larger predators. Coyotes—or, according to the research, dingoes—may do a far better job than humans of keeping feral cats in check, and without the ethical baggage. This approach "does not attempt to remove death from life," explains Arian Wallach, an ecologist at Charles Darwin University, "but it does change the relationship we have with the living."

In other words, if you want to kill a feral cat, get a wild dog.

In a study in the journal *Trends in Ecology and Evolution*, Wallach and a team looked at how large predators regulate so-called novel ecosystems, which is what ecologists call Anthropocene mashups of native and human-introduced species. Most conservationists consider novel ecosystems to be troubled, prone to domination by invasive species and boom-and-bust population cycles.

That's particularly problematic in Australia, which is basically an entire continent's worth of novel ecosystems plagued by cane toads, red foxes, and feral cats. Since Europeans arrived, more than

10 percent of Australia's mammals have gone extinct, and another 20 percent are threatened. Conservationists have responded with mostly ineffective trapping and poisoning campaigns; though controversial, Down Under people have less discomfort at killing cats than in North America, and Australia's government is now drafting a national feral cat control program.

Of course, the wild dogs are seen as a problem too. Dingoes, medium-sized canines that arrived several thousand years ago with Pacific Islander seafarers, are blamed for eating livestock (though they also help ranchers by eating kangaroos who compete with cattle for forage). Dingo extermination efforts have been patchy, though, which has created a sort of real-world experimental setup. In a nutshell: more dingoes, less cats, and vice versa. "In our current study site, we have about twice as many cats where dingoes are scarce," says Wallach.

That's no surprise. A 2012 study of interactions between dingoes and feral cats at nine study sites across Australia found much the same thing. "It may be that predators killing or scaring off other predators is our best bet," says that study's lead author, Euan Ritchie, an ecologist at Deakin University. "In a sense this is a positive, as it would see us working with nature rather than against it."

Ecologists know a lot less about feral cats in North America, but Ritchie and Wallach say similar dynamics likely hold, except with the dingoes' role played by coyotes. Key to the interaction, they say, isn't simply that dingoes and coyotes eat feral cats—though they do—but how their presence changes the cats' habits. Wary of being eaten, cats spend less time hunting and consume less prey. They might reproduce at lower rates or even move out altogether.

That's exactly what Stan Gehrt, who studies a thriving population of urban coyotes in the Chicago area, has found. "Where coyote densities were highest, we found that coyotes would kill a few cats, but not very many," explains Gehrt. "The reason is that cats were strongly avoiding those areas." Instead the cats moved into the city's most densely populated areas, sparing the green spaces and nature preserves where feral cat predation is the biggest problem. "It's a win-win," says Gehrt. "The cat people get a lot of cats who are still alive, and the coyotes are doing this strong ecological service. Everyone's happy." (Except maybe the cats.)

Gehrt's experience also suggests an avenue of redemption for coyotes, a species often treated with disregard and loathing. The conservation group Project Coyote estimates that people shoot, trap, or poison five hundred thousand a year in the United States.

Of course, it's possible that cat sympathizers won't be entirely happy with this solution. And when cats move deeper into cities, they're killing animals who might otherwise have been spared: pigeons and rats rather than songbirds and adorable critters, but creatures nonetheless deserving of respect. At least that's natural, though. "Feral cats are a part of their ecosystem," says Marc Bekoff, author of *Rewilding Our Hearts*. "But there's a really big difference between me poisoning a cat, and a coyote eating a cat. The coyote has to eat, but people don't have to kill feral cats." In ecosystems transformed by humans, it's still more ethical to let wild creatures find their own destinies.

SHOULD ANIMALS HAVE A RIGHT TO PRIVACY?

Last August a male bottlenose dolphin swam into a tributary of New Jersey's Raritan River. It was an unusual place for him to be. Bottlenose dolphins are saltwater creatures; they generally avoid rivers unless they're old and sick, as was this dolphin. After trying unsuccessfully to guide him back to sea, wildlife rescuers decided that saving the dolphin was impossible. Rather than prolonging his misery they euthanized him.

There the sad tale might have ended but for a journalist curious about why the dolphin died. She filed a public-records information request for the autopsy results. New Jersey's Department of Agriculture declined to share them, citing the dolphin's right to medical privacy. This was deeply weird. Although a great many people consider animals to be persons—not the same as us, necessarily, but obviously thinking and feeling beings—no governmental institution has ever formally recognized personhood outside of *Homo sapiens*.

In legal terms, personhood means an ability to possess rights. Courts in the United States have so far rejected every such claim, including high-profile lawsuits filed in neighboring New York State by the Nonhuman Rights Project, which argued that four chimps held in miserable conditions have a right to live somewhere decent. Even that bare-minimum request was too great, yet here the state of New Jersey seemed to affirm dolphins as people.

Some observers wondered if it might set a precedent. The short answer: almost certainly not. New Jersey did technically treat the dolphin as a person, but apparently department personnel just didn't think it through. "New Jersey in particular has shown very little interest in pushing its animal legal envelope," said the animal-law professor David Cassuto when I asked for his thoughts, and indeed the state quickly said it made a mistake. It's hard to imagine that loophole surviving a legal challenge; it would be revoked as soon as it hit a judge's desk.

That's probably for the best. A common rebuttal to giving legal rights to chimps is that animal rights claims would multiply and become impossibly messy. One can certainly imagine how a right to animal medical privacy could get frivolous. It could also be abused: "All the pigs in our factory died of this weird new flu, but the Cheap Bacon Megacorporation CEO feels it's important to respect their privacy at this difficult time."

That said, the notion that animals ought to have some sort of privacy rights isn't crazy. In some ways, society is already looking in that direction. The primatologist Catherine Hobaiter, at the University of St. Andrews, notes that human researchers make distinctions between public and private space. People can expect to be observed at a baseball game, but not in their bedrooms. She wonders if it's possible to extend that consideration to the wild chimps she studies.

David Favre, an animal-law scholar at Michigan State University, voiced a resonant note when I asked him about these ideas. "I'd want to think of it in the sense of a child," he said. "Even though they may not have a specific understanding of the idea of privacy, when do we know it's in their better interest to have privacy to protect them from the big bad world around us?" He mentioned the Detroit Zoo's installation of private spaces in their chimp habitat. "If you want to take into account their psychological well-being, it could entail the idea of not always having humans look at you."

Of course, chimps are extremely intelligent. It's perhaps easier to entertain privacy for them, humanity's closest living relative, than for a salamander or groundhog. What about other animals? Favre thinks it would be something to take on a species-by-species basis.

At the very least, we shouldn't just assume it's our Darwin-given privilege to barge into their lives with cameras and data-gathering equipment whenever we feel like it. Which is something to consider, given our extraordinary powers of intrusion into both animal and human lives.

Camera traps, GPS collars, landscape-scale acoustic monitoring, wildlife drones: we collect more animal information that ever. Here we should expand privacy beyond its psychological condition—not living under someone else's intrusive eye—and privacy as protection against the consequences of information gathering.

Although nonhunting-related surveillance of animals is usually done for the greater good of populations and species, there's growing criticism of methods that harm individuals in the process. Witness the outrage over a tracking collar that appeared to choke a polar bear (which may or may not have been the case, but it's not an unusual occurrence). In that context, privacy is intertwined with physical protection.

Related issues also emerged in the infamous "monkey selfie" lawsuit, decided earlier this month when a federal US court ruled that a monkey who'd used a photographer's camera didn't own the resulting snapshots, which went viral. The notion of a monkey being entitled to copyright protection earned some ridicule, yet the underlying tensions have crossed the mind of many wildlife photographers, including myself: Who benefits from my photographs? What am I imposing on the subjects? What's in it for them?

These questions might sound academic, but if photography affects an animal's fate, they're not. I stopped putting out camera traps after someone stole a memory card from one I'd installed near a hard-to-find animal trail used by the area's small mammals. My camera revealed their otherwise hidden route; that knowledge could be used to harm them.

These issues are especially relevant now. The ubiquity of digital recording and distribution tools, and the Internet-winning delight people take in watching animals, means that ours is a society in which just about everyone produces or consumes animal images. We're all implicated in the consequences of their production. The creatures depicted are not generic stand-ins. They're individuals making their way in the world, just like us.

How might animal privacy become a legal right rather than a cultural custom? Should it? I don't know. Those are complicated questions. But I do know that, a few days ago, walking in a park at dusk, I saw two raccoons curled up together on a branch atop an old, hollow tree that was probably their home. By all appearances it was a quiet, intimate moment, perhaps even a private one. I stopped to take photos. Through the lens I saw them looking at me. In retrospect, I wish I'd kept my camera in its bag.

WHEN CLIMATE CHANGE BLINDS US

In the Great Basin desert of the western United States, not far from the Great Salt Lake, is a kind of time machine. Homestead Cave has been inhabited for the past thirteen thousand years by successive generations of owls, beneath whose roosts accumulated millennia-deep piles of undigested fur and bone. By examining these piles, researchers have been able to reconstruct the region's ecological history. It contains a very timely lesson.

Those thirteen thousand years spanned some profound environmental upheavals. Indeed, the cave opened when Lake Bonneville, a vast prehistoric water body that covered much of the region, receded at the last ice age's end, and the Great Basin shifted from rainfall-rich coolness to its present hot, dry state. Yet despite these changes, life was pretty stable. Different species flourished at different times, but the total amount of biological energy—a metric used by ecologists to describe all the metabolic activity in an ecosystem—remained steady.

About a century ago, though, all that changed. There's now about 20 percent less biological energy flowing through the Great Basin than at the twentieth century's beginning. To put it another way: life's richness contracted by one-fifth in an eyeblink of geological time. The culprit? Not climate change, as one might expect, but human activity, in particular the spread of invasive nonnative grasses that flourish in disturbed areas and have little nutritional value, sustaining less life than would the native plants they've displaced.

I find myself thinking often of the parable of Homestead Cave, as I've come to call it. It underscores how resilient nature can be, and also the enormity of human impacts, which in this case dwarfed the transition to an entirely new climate state. The latter point, I fear, is too often overlooked these days, obscured by a fixation on climate change as Earth's great ecological problem.

Make no mistake: climate change is a huge, desperately important issue. And it feels strange, if not downright traitorous, to raise concerns about the attention it receives. The parable of Homestead Cave is no license to shirk climate duties on the assumption that nature will adapt, or to imagine that a rapidly warming, weather-extremed Earth won't be calamitous for nonhuman life. It will be. But so is a great deal else that we do. Paying attention to climate change and to other human impacts shouldn't be a zero-sum game, but it too often seems that way.

Witness the reception given to another study recently published in *Science* on Atlantic cod in the Gulf of Maine. A once-great fishery that's now a shadow of its historical abundance, cod stocks have failed to recover despite strict fishing restrictions enacted in 2010. Fast-rising water temperatures, said the researchers, seem to have changed Gulf of Maine food webs, depleting the cod's prey and slowing their recovery.

To be sure, the researchers noted a history of chronic overfishing, and made a nuanced point that fisheries management needs to incorporate environmental factors. But that nuance was lost in the flood of media attention to the study. The public narrative was exclusively about climate. "Climate Change Fuels Cod Collapse," proclaimed *The Boston Globe* last month; "Why Is It So Hard to Save Gulf of Maine Cod? They're in Hot Water," headlined NPR; and on and on, with barely a mention of the ecological upheavals that preceded the collapse.

These were enormous. Cod were brutally overfished since the beginning of the twentieth century. Bottom-trawling fishing methods destroyed many of their seafloor spawning grounds, and also the habitat of their prey. Before that, dams built on almost every coastal stream and river reduced once-vast populations of migratory fish such as Atlantic salmon and shad, billions of which had historically nourished cod, by at least 95 percent.

Not only had cod collapsed before the Gulf of Maine started to warm, but the entire ecosystem was also dramatically altered. Any sense of that was entirely absent from the study's coverage. Instead, the cod disaster was portrayed—ahistorically, misleadingly, and ultimately counterproductively—as a climate issue.

This wasn't the researchers' fault. It was the media's and the public's habit of mind—also visible earlier this year when Pope Francis released his encyclical *"Laudato Si": On Care for Our Common Home*. Promptly dubbed his "climate encyclical," coverage at first concentrated on the pope's call for action on climate change. In fact, the encyclical was so much more.

It was a full-throated denunciation of the ideology of unlimited economic growth, a lamentation of environmental degradation, a call for people to respect both ecosystems and individual animals: a radical, existentially challenging work, of which climate was just one facet. Yet it took months after its release, when critiques from environmental scholars and theologians supplanted the initial media coverage, for the climate frame to expand.

What explains this climate-first habit? Why do the effects of climate change on bird habitat, or on forest fires in the western United States, get headlines, whereas catastrophic fires set by Indonesian palm-oil producers merit a relative peep?

Perhaps it's some compensation for climate change having been treated with such a terrible lack of urgency in the United States and elsewhere. That would be understandable, though personally I'm not convinced. I think something else is at work: a certain lack of appreciation for just how big-footed 7.3 billion humans are, and also the fact that climate change is, in a sense, an easy out.

Reducing humanity's carbon pollution will certainly be logistically difficult, but its roots are essentially blameless—by the time climate change was a problem, nations had built their economies on cheap fossil fuel—and conceptually simple: pollute less. It's comforting to think that, if humanity can fix Earth's climate, nature's problems will be also be solved.

But that's not the case. It's all too easy to imagine a future in which humanity has averted the worst of climate change but nature is woefully diminished. The sixth great extinction won't be averted just because atmospheric CO2 levels fall below 350 parts per million. The United Nations climate conference has just concluded with an agreement that should help put humanity on a trajectory of climate sanity—but to protect the living world, we'll need to do much more.

TO BRING BACK EXTINCT SPECIES, WE'LL NEED TO CHANGE OUR OWN

The last passenger pigeon died just over a century ago, though they've lived on as symbols—of extinction's awful finality, and also of a human carelessness so immense that it could exterminate without really trying what was the most populous bird in North America. Centennials being a form of ritual, much has been written about them recently: about flocks a mile wide turning midday skies black and taking days to pass, descending on Eastern forests in a storm of life.

Their sheer, vanished ubiquity is evident in the list of places named after passenger pigeons, which can be found just about everywhere east of the Mississippi and below the Arctic. In New York, where I live, there's a Pigeon Mountain, Pigeon Creek, Pigeon Lake, two Pigeon Brooks, and no fewer than four Pigeon Hills. There's also a Pigeon Valley Cemetery and a Pigeon Valley Road, though the valley itself now goes by another name.

Yet to some people, the passenger pigeon's story doesn't need to have an unhappy ending; they might fly yet again, resurrected by biotechnology. They might become "de-extinct."

The notion was unveiled last spring at a daylong TEDx event held in Washington, DC, at the headquarters of the National Geographic Society, which dedicated an issue of its flagship magazine to the topic. On the issue's cover a parade of vanished animals poured forth from a test tube: a woolly mammoth and a saber-toothed cat, a giant sloth and a Tasmanian tiger, a giant moa and, of course, a pair of passenger pigeons.

"Do you want extinct species back?" said Stewart Brand, environmentalist and founder of Revive & Restore, a group of researchers and enthusiasts who organized the TEDx panel and coordinate de-extinction research. "Humans have made a huge hole in nature in the last 10,000 years. We have the ability now, and maybe the moral obligation, to repair some of the damage."

Those themes played out throughout the day, and throughout the de-extinction discourse: Bringing back. Righting wrongs. Re-creating. Resurrecting. And though a host of technical hurdles, described in depth by Carl Zimmer and Brian Switek in *National Geographic* and on the website, would first need to be overcome, it's at least possible to imagine de-extinction happening. We then run into a fundamental question: What do we mean by bringing a species back?

For one thing, the idea of "species" can overshadow the lives and experiences of individual animals. At one point during Brand's talk, he announced to applause the first de-extinction success: a cloned Pyrenaean ibex. Born with grossly deformed lungs, the baby goat lived for just ten minutes. The applause quickly subsided, and though the talk moved on to happier possibilities, I couldn't shake the thought of that poor goat, called into being only to suffer for the ideal of a species.

Assuming, though, that the reproductive technicalities are worked out painlessly, or that the pain is an acceptable cost: What, really, is a species? It's theoretically possible that a population of de-extinct animal could contain the entirety of its ancestral genome: a literal, physical embodiment of the lost species, appearing before our eyes as if pulled from history.

Yet those animals would arguably not be truly de-extinct. Whatever knowledge and habits once passed from bird to bird, flock to flock, anything not narrowly contained in genetic programming, would be lost in translation. Eventually, the engineered birds' descendants would acquire their own habits and knowledge. But they should be regarded that way: not de-extinct travelers from pre-industrial times, but creatures reinterpreted for the twenty-first century. And for that to happen would require, beyond all the challenges of DNA reconstruction and genome assembly, a feat of cultural engineering.

Critics of passenger pigeon de-extinction have rightly noted that the birds were not merely overhunted to oblivion. They lost their habitat, too: vast forests of oak and chestnut and beech were cut down and replaced by farms and cities. Some of those forests have returned, but a new generation of passenger pigeons would inevitably forage in farms and gardens. Would we be willing to share with them?

What of other de-extinction candidates? Would we protect estuaries from development for the eskimo curlew's sake, when we don't now protect them for endangered red knots? Ask timber companies to quit logging bottomland hardwood forests out of consideration for ivory-billed woodpeckers? Stop overfishing in the North Pacific so that Steller's sea cows have enough to eat?

It's difficult to imagine. All over the world, animals are dying out—either because they're worth more dead than alive, at least to some people, or because we don't want to share landscapes and natural resources with them. Wild-animal populations have fallen by 50 percent since the 1970s; not only are species going extinct, but there's also simply less of what remains.

Last month a northern white rhinoceros died at the San Diego Zoo. There are now just five left in the world. They do not appear to breed in captivity, and the possibility of using de-extinction techniques to save them has been discussed. In principle, I think it's worth trying—and, for the record, I also like the thought of engineering twenty-first-century versions of ivory-billed woodpeckers and eskimo curlews. It wouldn't be resurrection, or bringing them back, or righting a wrong, but it would be better than not doing it.

Now, however, it's the ghost of the northern white rhino, not the passenger pigeon, that haunts us. If we couldn't be bothered to save those rhinos from extinction the first time around, will we make room for them in the future? I hope so. But if we do, it won't be because of our technological prowess. It'll be because we had a change of heart.

SEPTEMBER 11, FALL MIGRATION, AND OCCUPY WALL STREET

On my way to Ground Zero on the tenth anniversary of September 11, 2001, I stopped for a slice a pizza and to clear my head. The previous week had been a somber one; every anniversary recalls the past, but some make you reflect on what's happened since, and a cloud hung over the intervening years. The nation felt like a different, far darker place than before that fateful morning.

Of course, it's easy to mythologize the past. Even the weather of 9/11, an archetypally perfect fall morning, takes on metaphorical overtones: a time of innocence and bounty, golden and pure, as yet untouched by shadow. Through the lens of memory, the United States was running a surplus, the economy was strong, things were good.

Of course they were not. A year before, the dot-com bubble burst, and with it the fantasy of economic security in an information age. A few months earlier, the Enron scandal surfaced—a Byzantine mix of accounting fraud, rigged markets, political corruption, ill-conceived deregulation, and greed and meanness and outright theft—perpetrated by people who preached the virtues of free markets and loaned the president their corporate jet.

Enron, we learned in years to come, wasn't an exception. It was a business model for big capitalism in the early twenty-first century. The same basic blueprint could be read in the financial meltdown of 2008 when investment bankers—who rewrote laws that once restrained them, pushed high-interest mortgages at the peak of a real estate bubble, bet trillions of dollars that mortgages would be paid even when they obviously wouldn't, then tried to hide these facts—crippled the economies of North America and Western Europe, and very nearly took down the world.

The consequences were quite different for poor and middle-class people than for hedge fund managers and investment bankers. Within a few years, as unemployment soared and cities went bankrupt, the people most responsible for the crisis were even wealthier than before. And between Enron's stock plunge and Lehman's bankruptcy we'd had two disastrous wars, state-sanctioned torture and surveillance, and the body politic's split into alternate partisan universes. Pervading it all was a sense of inescapability. Around and around we went, a society spiraling downward and unable to change course.

I jotted down my thoughts, finished eating, and walked to Ground Zero. There I said a prayer for the departed—I don't believe in God, but sometimes one just prays—and continued to my evening's destination, the Tribute in Lights, which is projected above lower Manhattan each 9/11 night. You've probably seen the tribute, or pictures of it, twin electric blue beams that disappear in the heavens and can be seen from 60 miles away. It is beautiful and utterly haunting and simply immense. For one night, it turns the rest New York's fabled skyline into a row of votive candles.

The year before, I'd seen the tribute from Governor's Island, just below Manhattan in New York Bay. I'd gone there for a concert, and during the opening act people drifted to the shoreline, where they looked at the tribute in wonder and confusion. There was something unusual about the beams: Sparkling white points of light spiraled slowly inside them, hundreds if not thousands, almost like confetti, but confetti wouldn't have been visible from that distance. It also wouldn't have risen. A few people said the lights made them think of souls.

The next day I learned from a friend that the lights had been birds. New York City sits directly in the Atlantic Flyway, the easternmost of North America's four great migration routes. Each fall, millions of birds fly down the Atlantic coast, a stream of energy and life stretching from Greenland in the summer to Tierra del Fuego, the southern tip of South America, in winter. Along the way the birds funnel down the Hudson River Valley, passing mostly unnoticed above the city that never sleeps.

Scientists aren't precisely sure how birds navigate their miraculous passage, but the general mechanisms are understood. They sense Earth's geomagnetic field, which provides a frame of reference calibrated by the light of stars, sun, and moon. Under certain conditions, however, such as moonless, overcast nights when the brightest lights are man-made, these biological compasses spin awry. Birds fly in circles until dropping from exhaustion onto sidewalks or stoops, or escape so drained as to die later in their journey.

September 11, 2010 had been one such night. The waxing moon was a thin, dim crescent. Clouds covered lower Manhattan. Birds had also gathered for days in wetlands north of the city, grounded by storms that blew against them, but finally the winds shifted to the south. In a tailwind flood the birds were released. The brightest light in the region came from the Tribute in Lights, projected by eighty-eight 7,000-watt xenon searchlights into a dull dark sky.

When I called New York City's chapter of the Audubon Society, I learned that more than ten thousand birds—yellow warblers on their way to Central America, redstarts headed to Mexico, probably tanagers and thrushes and orioles, too—were pulled in over the night's course. Five times Audubon volunteers briefly shuttered the spotlights, giving circling birds a chance to escape.

It seemed a noble thing to do, keeping our memorial to tragically lost life from accidentally taking lives; and so, for the tenth anniversary of 9/11, wanting to honor the day with more than remembrance, I volunteered, arriving just before dusk at the rooftop parking garage where the tribute's spotlights are installed.

Night fell. The sky over New Jersey turned from blue to purple to black. The lights hummed. Audubon volunteers lay on their backs, staring into the beams and trying to count the birds. There weren't many. Previous nights had favored flight, preventing the buildup seen a year earlier. Except for a few wispy clouds, the sky was clear, and the gibbous moon would soon be full. There seemed to be more people than birds: family members still grieving, tourists posing, a British man with a burn-scarred face who'd been installing floors at the World Trade Center on 9/11 and who mourned the Muslim lives lost since.

Only once, when clouds covered the moon a few hours after midnight, did birds enter the beams in significant numbers. The clouds soon blew away. The birds followed. As dawn approached, the beams were empty. Six days later, the first protesters arrived just down the block, at Zuccotti Park. Occupy Wall Street had begun.

MAKING SENSE OF 7 BILLION PEOPLE

On the last day of October 2011, the global population of an upstart branch of the primate order reached 7 billion.

What does it mean?

In itself, not much: 7 billion is just a one-digit flicker from 6,999,999,999. But the number carries a deep existential weight, symbolizing themes central to humanity's relationship with the rest of life on Earth.

For context, let's consider a few other numbers. The first: 10,000. That's approximately how many *Homo sapiens* existed 200,000 years ago, the date at which scientists mark the divergence of our species from the rest of *Homo genus*, of which we are the sole survivors.

From those humble origins, humans—thanks to our smarts, long-distance running skills, verbal ability, and skill with plants—proliferated at an almost inconceivable rate.

Some may note that, in a big-picture biological sense, humanity has rivals: in total biomass, ants weigh as much as we do, oceanic krill weigh more than both of us combined, and bacteria dwarf us all. Those are interesting factoids, but they belie a larger point.

Ants and krill and bacteria occupy an entirely different ecological level. A more appropriate comparison can be made between humans and other apex predators, which is precisely the ecological role humans evolved to play, and which—beneath our civilized veneer—we still are.

According to a back-of-the-envelope calculation, there are about 1.7 million other top-level, land-dwelling, mammalian predators on Earth. Put another way: For every nonhuman mammal sharing our niche, there are more than 4,000 of us.

In short, humans are Earth's great omnivore, and our omnivorous nature can only be understood at global scales. Scientists estimate that 83 percent of the terrestrial biosphere is under direct human influence. Crops cover some 12 percent of Earth's land surface, and account for more than one-third of terrestrial biomass. One-third of all available fresh water is diverted to human use.

Altogether, roughly 20 percent of Earth's net terrestrial primary production, the sheer volume of life produced on land on this planet every year, is harvested for human purposes—and, to return to the comparative factoids, it's all for a species that accounts for .00018 percent of Earth's nonmarine biomass.

We are the .00018 percent, and we use 20 percent. The purpose of that number isn't to induce guilt, or to blame humanity. The point of that number is perspective. At this snapshot in life's history, at—per the insights of James C. Rettie, who imagined life on Earth as a yearlong movie—a few minutes after 11:45 p.m. on December 31 we are big. Very big.

However, it must be noted that, as we've become big, much of life had to get out of the way. When modern *Homo sapiens* started scrambling out of East Africa, the average extinction rate of other mammals was, in scientific terms, 1 per million species years. It's 100 times that now, a number that threatens to make nonhuman life on Earth collapse.

In regard to that number, environmentalists usually say that humanity's fate depends on the life around us. That's debatable. Humans are adaptable and perfectly capable of living in squalor, without clean air or clean water or birds in the trees. If not, there wouldn't be 7 billion of us. Conservation is a moral question, and probably not a utilitarian imperative.

But the fact remains that, for all of humanity to experience a material standard of living now enjoyed by a tiny fraction, we'd need four more Earths. It's just not possible. And that, in the end, is the significance of 7 billion. It's a challenge.

In just a few minutes of evolutionary time, humanity has become a force to be measured in terms of the entirety of life itself. How do we, the God species, want to live? For an answer, check back at 8 billion.

REFERENCES

Introduction

Abram, David. "On Being Human in a More-Than-Human World." Center for Humans and Nature. http://www.humansandnature.org/to-be-human-david-abram.

Hillis, David. "Tree of Life." http://www.zo.utexas.edu/faculty/antisense/downloadfilestol.html.

Keim, Brandon. "Animal Kingdom Gets a New Root." *WIRED*, January 26, 2009. http://www.wired.com/2009/01/urmetazoan/.

——. "Crazy Sex Trick Fuels All-Male Clam Species." *WIRED*, May 23, 2011. http://www.wired.com/2011/05/clam-cloning/.

Pietsch, Theodore W. *Trees of Life: A Visual History of Evolution.* Baltimore: Johns Hopkins University Press, 2012.

I. Dynamics

Organized Chaos Makes the Beauty of a Butterfly

Feinberg, Andrew P., and Rafael A. Irizarry. "Stochastic Epigenetic Variation as a Driving Force of Development, Evolutionary Adaptation, and Disease." *Proceedings of the National Academy of Sciences* 107 (2010): 1757–64. doi:10.1073/pnas.0906183107.

Duret, Laurent. "Mutation Patterns in the Human Genome: More Variable Than Expected." *PLoS Biology* 7 (2009): e1000028. doi:10.1371/journal.pbio.1000028.

Frederick, Carl. "A Universe Made of Tiny, Random Chunks." *Nautilus,* June 6, 2013. http://nautil.us/issue/2/uncertainty/a-universe-made-of-tiny-random-chunks.

Lambert, Neill, et al. "Quantum Biology." *Nature Physics* 9 (2013): 10–18. doi:10.1038/nphys2474.

Losick, Richard, and Claude Desplan. "Stochasticity and Cell Fate." *Science* 5872 (2008): 65–68. doi:10.1126/science.1147888.

"Powers of Ten and the Relative Size of Things in the Universe." Eames Office. http://www.eamesoffice.com/the-work/powers-of-ten/.

Raj, Arjun, and Alexander van Oudenaarden. "Nature, Nurture, or Chance: Stochastic Gene Expression and Its Consequences." *Cell* 135 (2008): 216–226. doi:10.1016/j.cell.2008.09.050.

Whitford, Paul, Karissa Y. Sanbonmatsu, and José N. Onuchic. "Biomolecular Dynamics: Order-Disorder Transitions and Energy Landscapes." *Reports on Progress in Physics* 75 (2012): 076601. doi:10.1088/0034-4885/75/7/076601.

Whitford, Paul. "Disorder Guides Protein Function." *Proceedings of the National Academy of Sciences* 110 (2013): 7114–15. doi:10.1073/pnas.1305236110.

Chickadees, Mutations, and the Thermodynamics of Life

Gray, Suzanne M., and Jeffrey S. McKinnon. "Linking Color Polymorphism Maintenance and Speciation." *Trends in Ecology and Evolution* 22 (2007): 71–79. doi: 10.1016/j.tree.2006.10.005.

Keim, Brandon. "Something Other Than Adaptation Could Be Driving Evolution." *WIRED*, March 28, 2013. http://www.wired.com/2013/03/neutral-biodiversity/.

Leu, Kevin, et al. "The Prebiotic Evolutionary Advantage of Transferring Genetic Information from RNA to DNA." *Nucleic Acids Research* 39 (2011): 8135–47. doi:10.1093/nar/gkr525.

Martins, Ayana, Marcus A. M. de Aguiar, and Yaneer Bar-Yam. "Evolution and Stability of Ring Species." *Proceedings of the National Academy of Sciences* 110 (2013) 5080–84. doi:10.1073/pnas.1217034110.

Masel, Joanna, and Mark L. Siegal. "Robustness: Mechanisms and Consequences." *Trends in Genetics* 25 (2009): 395–403. doi:10.1016/j.tig.2009.07.005.

O'Dwyer, James P., and Ryan Chisholm. "A Mean Field Model for Competition: From Neutral Ecology to the Red Queen." *Ecology Letters* 17 (2014): 961–69. doi:10.1111/ele.12299.

Toussaint, Olivier, and Eric D. Schneider. "The Thermodynamics and Evolution of Complexity in Biological Systems." *Comparative Biochemistry and Physiology Part A: Molecular & Integrative Physiology* 120 (1998): 3–9. doi:10.1016/S1095-6433(98)10002-8.

The Photosynthetic Salamander

Graham, Erin R., Scott A. Fay, Adam Davey, and Robert W. Sanders. "Intracapsular Algae Provide Fixed Carbon to Developing Embryos of the Salamander *Ambystoma maculatum*." *Journal of Experimental Biology* 216 (2013): 452–59. doi:10.1242/jeb.076711.

Kerney, Ryan, et al. "Intracellular Invasion of Green Algae in a Salamander Host." *Proceedings of the National Academy of Sciences* 108 (2011): 6497–6502. doi:10.1073/pnas.1018259108.

Olivier, Heather M., and Brad R. Moon. "The Effects of Atrazine on Spotted Salamander Embryos and Their Symbiotic Alga." *Ecotoxicology* 19 (2010): 654–61. doi:10.1007/s10646-009-0437-8.

Yong, Ed. "The Unique Merger That Made You (and Ewe, and Yew)." *Nautilus*, February 6, 2014. http://nautil.us/issue/10/mergers—acquisitions/the-unique-merger-that-made-you-and-ewe-and-yew.

Human Evolution Enters an Exciting New Phase

Fu, Wenqing, et al. "Analysis of 6,515 Exomes Reveals the Recent Origin of Most Human Protein-Coding Variants." *Nature* 493 (2013): 216–20. doi:10.1038/nature11690.

Griffiths, R. C., and Simon Tavaré. "The Age of a Mutation in a General Coalescent Tree." *Stochastic Models* 14 (1998): 273–95. doi:10.1080/15326349808807471.

Hawks, et al. "Recent acceleration of human adaptive evolution." *Proceedings of the National Academy of Sciences* 104 (2007): 20753–58. doi: 10.1073/pnas.0707650104

Hayden, et al. "Cryptic Genetic Variation Promotes Rapid Evolutionary Adaptation in an RNA Enzyme." *Nature* 474 (2011): 92–95. doi:10.1038/nature10083.

Keim, Brandon. "Humans Evolving More Rapidly Than Ever, Say Scientists." *WIRED*, December 12, 2007. http://www.wired.com/2007/12/humans-evolving/.

——. "Beyond the Genome." *WIRED*, October 7, 2009. http://www.wired.com/2009/10/beyond-the-genome/.

——. "Cryptic Mutations Could Be Evolution's Hidden Fuel." *WIRED*, June 6, 2011. http://www.wired.com/2011/06/cryptic-variation/.

Keinan, Alon, and Andrew G. Clark. "Recent Explosive Human Population Growth Has Resulted in an Excess of Rare Genetic Variants." *Science* 336 (2012): 740–43. doi:10.1126/science.1217283.

Nelson, M. R., et al. "An Abundance of Rare Functional Variants in 202 Drug Target Genes Sequenced in 14,002 People." *Science* 337 (2012): 100–104. doi:10.1126/science.1217876.

Tennessen, Jacob A., et al. "Evolution and Functional Impact of Rare Coding Variation from Deep Sequencing of Human Exomes." *Science* 337 (2012): 64–69. doi:10.1126/science.1219240.

The 1,000 Genomes Project Consortium. "An Integrated Map of Genetic Variation from 1,092 Human Genomes." *Nature* 491 (2012): 56–65. doi:10.1038/nature11632.

Wade, Nicholas. "Many Rare Mutations May Underpin Diseases." *New York Times*, May 17, 2012. http://www.nytimes.com/2012/05/18/science/many-rare-mutations-may-underpin-diseases.html?_r=0.

"Parallel Universe" of Life Described Far beneath the Bottom of the Sea

Lever, Mark A., et al. "Evidence for Microbial Carbon and Sulfur Cycling in Deeply Buried Ridge Flank Basalt." *Science* 339 (2013): 1305–8. doi:10.1126/science.1229240.

Mason, Olivia U., et al. "First Investigation of the Microbiology of the Deepest Layer of Ocean Crust." *PLoS ONE* 5(2010): e15399. doi:10.1371/journal.pone.0015399.

At the Edge of Invasion, Possible New Rules for Evolution

Alford, Ross A., et al. "Comparisons through Time and Space Suggest Rapid Evolution of Dispersal Behaviour in an Invasive Species." *Wildlife Research* 36 (2009): 23–28. doi:10.1071/WR08021.
Gould, Stephen Jay, and Elisabeth Vrba. "Exaptation—A Missing Term in the Science of Form." *Paleobiology* 8 (1982): 4–15. doi:10.1017/S0094837300004310.
Shine, Richard, Gregory P. Brown, and Benjamin L. Phillips. "An Evolutionary Process That Assembles Phenotypes through Space Rather Than through Time." *Proceedings of the National Academy of Sciences* 108 (2011): 5708–11. doi:10.1073/pnas.1018989108.

A Mud-Loving, Iron-Lunged, Jelly-Eating Ecosystem Savior

Coll, Marta, Heike K. Lotze, and Tamara N. Romanuk. "Structural Degradation in Mediterranean Sea Food Webs: Testing Ecological Hypotheses Using Stochastic and Mass-Balance Modelling." *Ecosystems* 11 (2008): 939–60. doi:10.1007/s10021-008-9171-y.
Lejeusne, Christophe, et al. "Climate Change Effects on a Miniature Ocean: The Highly Diverse, Highly Impacted Mediterranean Sea." *Trends in Ecology and Evolution* 25 (2010): 250–60. doi:10.1016/j.tree.2009.10.009.
Utne-Palm, Anne C., et al. "Trophic Structure and Community Stability in an Overfished Ecosystem." *Science* 5989 (2010): 333–36. doi:10.1126/science.1190708.

Redeeming the Lamprey

Berman, Taylor. "Terrifying Sea Monster Found in New Jersey River." *Gawker*, February 26, 2013. http://gawker.com/5987185/terrifying-sea-monster-found-in-new-jersey-river.
Edelman, Adam. "Photos of 'Monster' Eel Fished out of New Jersey Waters Draw 1.2 million Views on Reddit." *New York Daily News*, February 25, 2013. http://www.nydailynews.com/news/national/photos-new-jersey-eel-wow-viewers-article-1.1273403.
"Friend Also Caught This Fishing in NJ." Reddit. February 15, 2013. https://www.reddit.com/r/WTF/comments/18ky4p/friend_also_caught_this_fishing_in_nj/.
Hogg, Robert Scott. "Fish Community Response to a Small-Stream Dam Removal in a Maine Coastal River Tributary." MS thesis, University of Maine, 2012.
Hogg, Robert, Stephen M. Coghlan Jr., Joseph Zydlewski and Kevin S. Simon. "Anadromous Sea Lampreys (*Petromyzon marinus*) are Ecosystem

Engineers in a Spawning Tributary." *Freshwater Biology* 59 (2014): 1294–1307. doi:10.1111/fwb.12349.

Hogg, Robert, Stephen M. Coghlan Jr., and Joseph Zydlewski. "Anadromous Sea Lampreys Recolonize a Maine Coastal River Tributary after Dam Removal." *Transactions of the American Fisheries Society* 142 (2013): 1381–94. doi:10.1080/00028487.2013.811103.

Keim, Brandon. "Return of the Ghost Fish." *On Earth*, November 12, 2013. http://archive.onearth.org/articles/2013/11/can-civilization-and-salmon-coexist-dam-good-question.

Ketola, et al. "Effects of Thiamine on Reproduction of Atlantic Salmon and a New Hypothesis for Their Extirpation in Lake Ontario." *Transactions of the American Fisheries Society* 129 (2000): 607–12. doi:10.1577/1548-8659(2000)129<0607:EOTORO>2.0.CO;2.

Naiman, H. George, Paul R. Bowser, Gregory A. Wooster, and Steven S. Hurst. "Pacific Salmon, Nutrients, and the Dynamics of Freshwater and Riparian Ecosystems." *Ecosystems* 5 (2002): 399–417. doi:10.1007/s10021-001-0083-3.

Nislow, Keith, and Boyd E. Kynard. "The Role of Anadromous Sea Lamprey in Nutrient and Material Transport between Marine and Freshwater Environments." *American Fisheries Society Symposium* 69 (2009): 485–94.

Waldman, John. *Running Silver: Restoring Atlantic Rivers and Their Great Fish Migrations*. Guilford, CT: Lyons Press, 2013.

Waldman, John R., Cheryl Grunwald, Nirmal K. Roy, and Isaac I. Wirgin. "Mitochondrial DNA Analysis Indicates Sea Lampreys Are Indigenous to Lake Ontario." *Transactions of the American Fisheries Society* 133 (2004): 950–60. doi:10.1577/T03-104.1.

Decoding Nature's Soundtrack

Farina, Almo, Emanuele Lattanzi, Rachele Malavasi, Nadia Pieretti, Luigi Piccioli. "Avian Soundscapes and Cognitive Landscapes: Theory, Application and Ecological Perspectives." *Landscape Ecology* 26 (2011): 1257–67 doi:10.1007/s10980-011-9617-z.

Farina, Almo, and Nadia Pieretti. "The Soundscape Ecology: A New Frontier of Landscape Research and Its Application to Islands and Coastal Systems." *Journal of Marine and Island Cultures* 1 (2012): 21–26. doi:10.1016/j.imic.2012.04.002.

——. "From Umwelt to Soundtope: An Epistemological Essay on Cognitive Ecology." *Biosemiotics* 7 (2013): 1–10. doi:10.1007/s12304-013-9191-7.

Krause, Bernie. *The Great Animal Orchestra: Finding the Origins of Music in the World's Wild Places*. New York: Little, Brown and Company, 2012.

——. "The Sound of a Damaged Habitat." *New York Times*, July 28, 2012. http://www.nytimes.com/2012/07/29/opinion/sunday/listen-to-the-soundscape.html.

Krause, Bernie, and Almo Farina. "Using Ecoacoustic Methods to Survey the Impacts of Climate Change on Biodiversity." *Biological Conservation* 195 (2016) 245–54. doi:10.1016/j.biocon.2016.01.013.

Krause, Bernie, Stuart H. Gage, and Wooyeong Joo. "Measuring and Interpreting the Temporal Variability in the Soundscape at Four Places in Sequoia National Park." *Landscape Ecology* 26 (2011): 1247–56. doi:10.1007/s10980-011-9639-6.

Pieretti, Nadia, and Almo Farina. "Application of a Recently Introduced Index for Acoustic Complexity to an Avian Soundscape with Traffic Noise." *The Journal of the Acoustical Society of America* 134 (2013): 891–900. doi:10.1121/1.4807812.

Pijanowski, Bryan C., et al. "Soundscape Ecology: The Science of Sound in the Landscape." *BiosScience* 61 (2011): 203–16. doi:10.1525/bio.2011.61.3.6.

——. "What Is Soundscape Ecology? An Introduction and Overview of an Emerging New Science." *Landscape Ecology* 26 (2011): 1213–32. doi:10.1007/s10980-011-9600-8.

Sueur, Jérôme, Sandrine Pavoine, Olivier Hamerlynck, Stéphanie Duvail. "Rapid Acoustic Survey for Biodiversity Appraisal." *PLoS ONE* 3 (2008): e4065. doi:10.1371/journal.pone.0004065.

II. Inner Lives

Being a Sandpiper

Allen, Timothy A., and Norbert J. Fortin. "The Evolution of Episodic Memory." *Proceedings of the National Academy of Sciences* 110, Supplement 2 (2013): 10379–86. doi:10.1073/pnas.1301199110.

Antunes, et al. "Individually Distinctive Acoustic Features in Sperm Whale Coda." *Animal Behaviour* 81 (2011): 723–30. doi:10.1016/j.anbehav.2010.12.019.

Ben-Ami Bartal, Inbal, Jean Decety, and Peggy Mason. "Helping a Cagemate in Need: Empathy and Pro-Social Behavior in Rats." *Science* 334 (2011): 1427–30. doi:10.1126/science.1210789.

Birkhead, Tim. *Bird Sense: What It's Like to Be a Bird*. New York: Walker Publishing, 2012.

"*Calidris pusilla* (Semipalmated Sandpiper)." The IUCN Red List of Threatened Species. http://www.iucnredlist.org/details/22693373/0.

"The Cambridge Declaration on Consciousness." Francis Crick Memorial Conference 2012: Consciousness in Human and Non-Human Animals. http://fcmconference.org/img/CambridgeDeclarationOnConsciousness.pdf.

Careau, V., D. Thomas, M. M. Humphries, and D. Réale. "Energy Metabolism and Animal Personality." *Oikos* 117 (2008): 641–53. doi:10.1111/j.0030-1299.2008.16513.x.

Carruthers, Peter. "Evolution of Working Memory." *Proceedings of the National Academy of Sciences* 110, Supplement 2 (2013): 10371–78. doi:10.1073/pnas.1301195110.

Darwin, Charles. *The Expression of the Emotions in Man and Animals*. London: John Murray, 1872.

Driedzic, William R., Heidi L. Crowe, Peter W. Hicklin, and Dawn H. Sephton. "Adaptations in Pectoralis Muscle, Heart Mass, and Energy

Metabolism during Premigratory Fattening in Semipalmated Sandpipers (*Calidris pusilla*)." *Canadian Journal of Zoology* 71(1993): 1602–8. doi:10.1139/z93-226.

Emery, Nathan J. "Cognitive Ornithology: The Evolution of Avian Intelligence." *Philosophical Transactions of the Royal Society B*. 361 (2006): 23–43. doi:10.1098/rstb.2005.1736.

Flynn, Laura, Erica Nol, and Yuri Zharikov. "Philopatry, Nest-Site Tenacity, and Mate Fidelity of Semipalmated Plovers." *Journal of Avian Biology* 30 (1999): 47–55. doi:10.2307/3677242.

Foucault, Michael. *The Order of Things*. New York: Pantheon Books, 1970.

Gratto, Cheri L., R. I. G. Morrison, and Fred Cooke. "Philopatry, Site Tenacity, and Mate Fidelity in the Semipalmated Sandpiper." *The Auk* 102 (1985): 16–24. doi:10.2307/4086818.

Gratto-Trevor, Cheri L., and Curt M. Vacek. "Longevity Record and Annual Adult Survival of Semipalmated Sandpipers." *The Wilson Bulletin* 113 (2001): 348–50. doi:http://dx.doi.org/10.1676/0043-5643(2001)113[0348:LRAAAS]2.0.CO;2.

Griffin, Donald R. *The Question of Animal Awareness: Evolutionary Continuity of Mental Experience*. New York: Rockefeller University Press, 1977.

Hicklin, Peter W., and John W. Chardine. "The Morphometrics of Migrant Semipalmated Sandpipers in the Bay of Fundy: Evidence for Declines in the Eastern Breeding Population." *Waterbirds* 35 (2012): 74–82. doi:10.1675/063.035.0108.

Jehl, Joseph R., Jr. "Coloniality, Mate Retention, and Nest-Site Characteristics in the Semipalmated Sandpiper." *The Wilson Journal of Ornithology* 118 (2006): 478–84. doi:10.1676/05-120.1.

——. "Disappearance of Breeding Semipalmated Sandpipers from Churchill, Manitoba: More Than a Local Phenomenon." *The Condor* 109 (2007): 351–60. doi:10.1650/0010-5422(2007)109[351:DOBSSF]2.0.CO;2.

Kaufman, Ken. "Lives of North American Birds." New York: Houghton Mifflin, 1996.

Mashour, George A., and Michael T. Alkire. "Evolution of Consciousness: Phylogeny, Ontogeny, and Emergence from General Anesthesia." *Proceedings of the National Academy of Sciences* 110, Supplement 2 (2013): 10357–64. doi:10.1073/pnas.1301188110.

McPhee, John. *The Founding Fish*. New York: Farrar, Straus and Giroux, 2002.

Mizrahi, et al. "Patterns of Corticosterone Secretion in Migrating Semipalmated Sandpipers at a Major Spring Stopover Site." *The Auk* 118 (2001): 79–91. doi:10.1642/0004-8038(2001)118[0079:POCSIM]2.0.CO;2.

Morell, Virginia. *Animal Wise: The Thoughts and Emotions of Our Fellow Creatures*. New York: Crown Publishers, 2013.

Morrison, R. I. Guy. "Dramatic Declines of Semipalmated Sandpipers on Their Major Wintering Areas in the Guianas, Northern South America." *Waterbirds* 35(2012): 120–34. doi:10.1675/063.035.0112.

Ostojic, et al. "Evidence Suggesting That Desire-State Attribution May Govern Food Sharing in Eurasian Jays." *Proceedings of the National Academy of Sciences* 110 (2013): 4123–28. doi:10.1073/pnas.1209926110.
Piersma, Theunis. "When a Year Takes 18 Months: Evidence for a Strong Circannual Clock in a Shorebird." *Science of Nature* 89 (2002): 278–89. doi: 10.1007/s00114-002-0325-z.
Prosek, James. *Trout: An Illustrated History.* New York: Knopf, 1996.
Prosek, James. *Ocean Fishes: Paintings of Saltwater Fish.* New York: Rizzoli, 2012.
"Sandpiper—Poem by Elizabeth Bishop." Poemhunter.com. http://www.poemhunter.com/poem/sandpiper/.
"Semipalmated Sandpiper." The Birds of North America Online. http://bna.birds.cornell.edu/bna/species/006/articles/introduction.
"Semipalmated Sandpiper." The Cornell Lab of Ornithology: All about Birds. https://www.allaboutbirds.org/guide/Semipalmated_Sandpiper/lifehistory.
Seyfarth, Robert M., and Dorothy L. Cheney. "Affiliation, Empathy, and the Origins of Theory of Mind." *Proceedings of the National Academy of Sciences* 110, Supplement 2 (2013): 10349–56. doi:10.1073/pnas.1301223110.
Shaw, Lyttle. "Pruning Names." *Cabinet* (Spring 2002). http://www.cabinetmagazine.org/issues/6/shaw.php.
Townsend, Charles Wendell. *The Birds of Essex County, Massachusetts.* Cambridge, MA: The Nuttall Ornithological Club, 1905.
van Zomeren, Koos. "Yes, That Was Peter." *NRC* (March 9, 2002). https://www.nrc.nl/nieuws/2002/03/09/ja-dat-was-peter-7580746-a1376005.

Monogamy Helps Geese Reduce Stress

Breuer, George. *Sociobiology and the Human Dimension.* Cambridge: Cambridge University Press, 1982.
Huber, Robert, and Michael Marys. "Male-Male Pairs in Greylag Geese (*Anser anser*)." *Journal of Ornithology* 134 (1993): 155–64. doi:10.1007/BF01640084.
Miyazaki, Takao, et al. "Relationship between Perceived Social Support and Immune Function." *Stress & Health* 19 (2003): 3–7. doi:10.1002/smi.950.
Wascher, Claudia A. F., Isabella B. R. Scheiber, Brigitte Weiß, and Kurt Kotrschal. "Heart Rate Responses to Agonistic Encounters in Greylag Geese, *Anser anser.*" *Animal Behaviour* 77 (2009): 955–61. doi:10.1016/j.anbehav.2009.01.013.
Wascher, Claudia A. F., Orlaith N. Fraser, and Kurt Kotrschal. "Heart Rate during Conflicts Predicts Post-Conflict Stress-Related Behavior in Greylag Geese." *PLoS ONE* 5 (2010): e15751. doi:10.1371/journal.pone.0015751.
Wascher, Claudia A. F., Brigitte Weiß, Walter Arnold, and Kurt Kotrschal. "Physiological Implications of Pair-Bond Status in Greylag Geese." *Biology Letters* 8 (2012): 347–50. doi:10.1098/rsbl.2011.0917.

What Pigeons Teach Us about Love

Bekoff, Marc. *Animal Passions and Beastly Virtues: Reflections on Redecorating Nature*. Philadelphia: Temple University Press, 2006.

——. *The Emotional Lives of Animals: A Leading Scientist Explores Animal Joy, Sorrow, and Empathy—and Why They Matter*. San Francisco: New World Library, 2007.

Black, Jeffrey M., ed. *Partnerships in Birds: The Study of Monogamy*. Oxford: Oxford University Press, 1996.

Chang Chi. "A Faithful Wife." In *Love and the Turning Year: One Hundred More Poems from the Chinese*, edited by Kenneth Rexroth, 82. New York: New Directions Books, 1970.

"Do Pigeons Mourn Their Dead?" Pigeon-Talk. http://www.pigeons.biz/forums/f27/do-pigeons-mourn-their-dead-21224.html.

Francesco, Franza, and Alba Cervone. "Neurobiology of Love." *Pschiatria Danubina* 26 (2014): 266–68. PMID: 25413551.

Freeman, Don. *Fly High, Fly Low*. New York: Viking Books for Young Readers, 2004.

Highsmith, Patricia. "Two Disagreeable Pigeons." In *Patricia Highsmith: Selected Novels and Short Stories*, edited by Joan Schenkar, 585–92. New York: W. W. Norton, 2011.

"How Can I Help a Grieving Pigeon?" Pigeon-Talk. http://www.pigeons.biz/forums/f27/how-can-i-help-a-grieving-pigeon-7705.html.

Klatt, James D., and James L. Goodson. "Oxytocin-Like Receptors Mediate Pair Bonding in a Socially Monogamous Songbird." *Proceedings of the Royal Society B*, 280 (2013): 20122396. doi:10.1098/rspb.2012.2396.

Masson, Jeffrey Moussaieff, and Susan McCarthy. *When Elephants Weep: The Emotional Lives of Animals*. New York: Delacorte Press, 1995.

Patel, Kruti K., and Courtney Siegel. "Genetic Monogamy in Captive Pigeons (*Columba livia*) Assessed by DNA Fingerprinting." *BIOS* 76 (2005): 97–101. doi:10.1893/0005-3155(2005)076[0097:RAGMIC]2.0.CO;2.

"Pigeons Moving on from Grief. . . ." Pigeon-Talk. http://www.pigeons.biz/forums/f27/pigeons-moving-on-from-grief-25520.html.

Rolston, Holmes, III. *A New Environmental Ethics: The Next Millennium for Life on Earth*. New York: Routledge, 2012.

Wascher, Claudia A. F., Brigitte Weiß, Walter Arnold, and Kurt Kotrschal. "Physiological Implications of Pair-Bond Status in Greylag Geese." *Biology Letters* 8 (2012): 347–50. doi:10.1098/rsbl.2011.0917.

Young, Kevin. "Ragtime." Poetry Society of America. https://www.poetrysociety.org/psa/poetry/poetry_in_motion/atlas/portland/ragtime/.

Young, Larry J. "The Neural Basis of Pair Bonding in a Monogamous Species: A Model for Understanding the Biological Basis of Human Behavior." In *Offspring: Human Fertility Behavior in Biodemographic Perspective*, edited by Kenneth W. Wachter and Rodolfo A. Bulatao, 91–103. Washington, DC: National Academies Press, 2003.

Zimmer, Carl. "Pigeons Get a New Look." *New York Times*, February 4, 2013. http://www.nytimes.com/2013/02/05/science/pigeons-a-darwin-favorite-carry-new-clues-to-evolution.html?pagewanted=all&_r=0.

Chimps and the Zen of Falling Water

Becker, Brett, and Courtney Kneipp. "Video of the Southern Resident Ceremony on Oct. 4, 2005." Beam Reach. http://www.beamreach.org/051/movie/ceremony/.

Bekoff, Marc. *The Emotional Lives of Animals: A Leading Scientist Explores Animal Joy, Sorrow, and Empathy—and Why They Matter.* San Francisco: New World Library, 2007.

Fisher, Christopher L. "Animals, Humans and X-men: Human Uniqueness and the Meaning of Personhood." *Theology and Science* 3 (2005): 291–314. doi:10.1080/14746700500317289.

Goldman, Jason G. "Death Rituals in the Animal Kingdom." *BBC Future*, September 19, 2012. http://www.bbc.com/future/story/20120919-respect-the-dead.

Harrod, James. "The Case for Chimpanzee Religion." *Journal for the Study of Religion, Nature and Culture* 8 (2014): 8–45. doi:0.1558/jsrnc.v8i1.8.

Legare, et al. "Imitative Flexibility and the Development of Cultural Learning." *Cognition* 142 (2015): 351–61. doi:10.1016/j.cognition.2015.05.020.

"List of Water Deities." Wikipedia. https://en.wikipedia.org/wiki/List_of_water_deities.

McGrew, William. *The Cultured Chimpanzee: Reflections on Cultural Primatology*. Cambridge: Cambridge University Press, 2004.

——. "Affidavit of William C. McGrew." http://www.nonhumanrightsproject.org/wp-content/uploads/2013/12/Suffolk-Ex.-10-McGrew-Affidavit-to-VP.pdf.

Perry, Susan. "Social Traditions and Social Learning in Capuchin Monkeys (*Cebus*)." *Philosophical Transactions of the Royal Society B*. 366 (2011): 988–96. doi:10.1098/rstb.2010.0317.

Pruetz, Jill D., and Thomas C. LaDuke. "Brief Communication: Reaction to Fire by Savanna Chimpanzees (*Pan troglodytes verus*) at Fongoli, Senegal: Conceptualization of 'Fire Behavior' and the Case for a Chimpanzee Model." *American Journal of Physical Anthropology* 141 (2009): 646–50. doi:10.1002/ajpa.21245.

"Waterfall Displays." The Jane Goodall Institute. https://vimeo.com/18404370.

How City Living Is Reshaping the Brains and Behavior of Urban Animals

"Conservation Genetics of Urban Stream Salamanders." Munshi-South Lab. http://nycevolution.org/research/conservation-genetics-of-urban-stream-salamanders/.

Evans, Jackson, Kyle Boudreau, and Jeremy Hyman. "Behavioural Syndromes in Urban and Rural Populations of Song Sparrows." *Ethology* 116 (2010): 588–95. doi:10.1111/j.1439-0310.2010.01771.x.

GrrlScientist. "American Crows: The Ultimate Angry Birds?" *The Guardian*, July 6, 2011. https://www.theguardian.com/science/punctuated-equilibrium/2011/jul/06/2df.

Harris, Stephen E., Jason Munshi-South, Craig Obergfell, and Rachel O'Neill. "Signatures of Rapid Evolution in Urban and Rural Transcriptomes of White-Footed Mice (*Peromyscus leucopus*) in the New York Metropolitan Area." *PLoS ONE* 8 (2013): e74938. doi:10.1371/journal.pone.0074938.

Keim, Brandon. "Why Some Wild Animals Are Becoming Nicer." *WIRED*, February 7, 2012. http://www.wired.com/2012/02/self-domestication/.

Miranda, Ana Catarina, Holger Schielzeth, Tanja Sonntag, and Jesko Partecke. "Urbanization and Its Effects on Personality Traits: A Result of Microevolution or Phenotypic Plasticity?" *Global Change Biology* 19 (2013): 2634–44. doi:10.1111/gcb.12258.

Partan, Sarah R., Andrew G. Fulmer, Maya A. M. Gounard, and Jake E. Redmond. "Multimodal Alarm Behavior in Urban and Rural Gray Squirrels Studied by Means of Observation and a Mechanical Robot." *Current Zoology* 56 (2010): 313–26.

Scales, Jennifer, Jeremy Hyman, and Melissa Hughes. "Behavioral Syndromes Break Down in Urban Song Sparrow Populations." *Ethology* 117 (2011): 887–95. doi:10.1111/j.1439-0310.2011.01943.x.

Snell-Rood, Emilie C., and Naomi Wick. "Anthropogenic Environments Exert Variable Selection on Cranial Capacity in Mammals." *Proceedings of the Royal Society B* 280 (2013): 20131384. doi:10.1098/rspb.2013.1384.

Reconsider the Rat: The New Science of a Reviled Rodent

Ben-Ami Bartal, Inbal, et al. "Helping a Cagemate in Need: Empathy and Pro-Social Behavior in Rats." *Science* 334 (2011): 1427–30. doi:10.1126/science.1210789.

——. "Pro-Social Behavior in Rats Is Modulated by Social Experience." *eLife* 3 (2014): e01385. doi:10.7554/eLife.01385.

Bellini, Jarrett. "Apparently This Matters: A Ghost Ship with Cannibal Rats." *CNN*, January 27, 2014. http://www.cnn.com/2014/01/24/tech/web/apparently-this-matters-lyubov-orlova-ghost-ship/.

C.N. Slobodchikoff, Bianca S. Perla, and Jennifer L. Verdolin. *Prairie Dogs: Communication and Community in an Animal Society*. Cambridge, MA: Harvard University Press, 2009.

Darwin, Charles. *The Descent of Man, and Selection in Relation to Sex*, vol. 1. New York: D. Appleton, 1872.

Decety, Jean, Greg J. Norman, Gary G. Berntson, and John T. Cacioppo. "A Neurobehavioral Evolutionary Perspective on the Mechanisms Underlying Empathy." *Progress in Neurobiology* 98 (2012): 38–48. doi:10.1016/j.pneurobio.2012.05.001.

de Waal, Frans. "Ants Are Not Rats." Frans de Waal—Public Page. https://www.facebook.com/permalink.php?story_fbid=365130943555593&id=99206759699.

——. "Putting the Altruism Back into Altruism: The Evolution of Empathy." *Annual Review of Psychology* 59 (2008): 279–300. doi:10.1146/annurev.psych.59.103006.093625.

Hecht, E. E., R. Patterson, and A. K. Barbey. "What Can Other Animals Tell Us about Human Social Cognition? An Evolutionary Perspective on Reflective and Reflexive Processing." *Frontiers in Human Neuroscience* 6 (2012): 224. doi:10.3389/fnhum.2012.00224.

Keim, Brandon. "Why Some Wild Animals Are Becoming Nicer." *WIRED*, February 7, 2012. http://www.wired.com/2012/02/self-domestication/.

——. "How City Living Is Reshaping the Brains and Behavior of Urban Animals." *WIRED*, August 22, 2013. http://www.wired.com/2013/08/urban-animal-brain-behavior-evolution/all/.

Pelz, Hans-Joachim et al. "The Genetic Basis of Resistance to Anticoagulants in Rodents." *Genetics* 170 (2005): 1839–47. doi:10.1534/genetics.104.040360.

"The Rodent Who Knew Too Much." *Science*, March 8, 2007. http://www.sciencemag.org/news/2007/03/rodent-who-knew-too-much.

Rodwell, James. *The Rat: Its History & Destructive Character.* London: G. Routledge & Co., 1858.

Snell-Rood, Emilie C., and Naomi Wick. "Anthropogenic Environments Exert Variable Selection on Cranial Capacity in Mammals." *Proceedings of the Royal Society B* 280 (2013): 20131384. doi:10.1098/rspb.2013.1384.

Sorensen, Eric. "The Animal Mind Reader." *Washington State Magazine*, Summer 2013. http://wsm.wsu.edu/s/index.php?id=1037.

Stone, Witmer, and William Everett Cram. *American Animals: A Popular Guide to the Mammals of North America North of Mexico, with Intimate Biographies of the Most Familiar Species.* New York: Doubleday, Page & Company, 1902.

Sullivan, Robert. *Rats: Observations on the History & Habitat of the City's Most Unwanted Inhabitants.* New York: Bloomsbury Publishing, 2004.

Vasconcelos, et al. "Pro-Sociality without Empathy." *Biology Letters* 8 (2012): 910–12. doi:10.1098/rsbl.2012.0554.

Monkeys See Selves in Mirror, Open a Barrel of Questions

Beran, Michael J., David J. Smith, Joshua S. Redford, and David A. Washburn. "Rhesus Macaques (*Macaca mulatta*) Monitor Uncertainty during Numerosity Judgments." *Journal of Experimental Psychology: Animal Behavior Processes* 32 (2006): 111–19. doi:10.1037/0097-7403.32.2.111.

Chang, Liangtang, Qin Fang, Shikun Zhang, Mu-ming Poo, Neng Gong. "Mirror-Induced Self-Directed Behaviors in Rhesus Monkeys after Visual-Somatosensory Training." *Current Biology* 25 (2015): 212–17. doi:10.1016/j.cub.2014.11.016.

Rajala, Abigail Z., Katharine R. Reininger, Kimberly M. Lancaster, Luis C. Populin. "Rhesus Monkeys (*Macaca mulatta*) Do Recognize Themselves in the Mirror: Implications for the Evolution of Self-Recognition." *PLoS ONE* 5 (2010): e12865. doi:10.1371/journal.pone.0012865.

The New Anthropomorphism

Ackerman, Jennifer. *The Genius of Birds*. New York: Penguin Press, 2016.
Balcombe, Jonathan. *What a Fish Knows: The Inner Lives of Our Underwater Cousins*. New York: Farrar, Straus and Giroux, 2016.
Barron, Andrew B., and Colin Klein. "What Insects Can Tell Us about the Origins of Consciousness." *Proceedings of the National Academy of Sciences* 113 (2016): 4900–4908. doi:10.1073/pnas.1520084113.
Ben-Ami Bartal, Inbal, Jean Decety, and Peggy Mason. "Helping a Cagemate in Need: Empathy and Pro-Social Behavior in Rats." *Science* 334 (2011): 1427–30. doi:10.1126/science.1210789.
Ben-Ami Bartal, et al. "Anxiolytic Treatment Impairs Helping Behavior in Rats." *Frontiers in Psychology* 7 (2016): 850. doi:10.3389/fpsyg.2016.00850.
Brown, Culum. "Fish Intelligence, Sentience and Ethics." *Animal Cognition* 18 (2015): 1–17. doi:10.1007/s10071-014-0761-0.
——. "Comparative Evolutionary Approach to Pain Perception in Fishes." *Animal Sentience* (2016): 2016.011.
Burghardt, Gordon. "Cognitive Ethology and Critical Anthromorphism: A Snake with Two Heads and Hognose Snakes That Play Dead." In *Cognitive Ethology: The Minds of Other Animals*, edited by C. A. Ristau, 53–90. San Francisco: Erlbaum, 1991.
——. "Amending Tinbergen: A Fifth Aim for Ethology." In *Anthropomorphism, Anecdotes, and Animals*, edited by R.W. Mitchell, N.S. Thompson, and H.L. Miles, 254–76. Albany, NY: SUNY Press, 1997.
——. "The Evolutionary Origins of Play Revisited: Lessons from Turtles." In *Animal Play: Evolutionary, Comparative and Ecological Perspectives*, edited by Marc Bekoff and John A. Byers, 1–26. Cambridge: Cambridge University Press, 1998.
——. "Mediating Claims through Critical Anthropomorphism." *Animal Sentience* (2016): 2016.024.
Corballis, Michael C. "Mental Time Travel: A Case for Evolutionary Continuity." *Trends in Cognitive Sciences* 17 (2013): 5–6. doi:10.1016/j.tics.2012.10.009.
Davis, Karen Marie, and Gordon Burghardt. "Turtles (*Pseudemys nelsoni*) Learn about Visual Cues Indicating Food from Experienced Turtles." *Journal of Comparative Psychology* 125 (2011): 404–10. doi:10.1037/a0024784.
——. "Long-Term Retention of Visual Tasks by Two Species of Emydid Turtles, *Pseudemys nelsoni* and *Trachemys scripta*." *Journal of Comparative Psychology* 126 (2012): 213–23. doi:10.1037/a0027827.
de Waal, Frans. *Are We Smart Enough to Know How Smart Animals Are?* New York: W. W. Norton, 2016.
Feeney, Miranda C., William A. Roberts, and David F. Sherry. "Black-Capped Chickadees *(Poecile atricapillus)* Anticipate Future Outcomes of Foraging Choices." *Journal of Experimental Psychology: Animal Behavior Processes* 37 (2011): 30–40. doi:10.1037/a0019908.

Franks, Becca, E. Tory Higgins, and Frances A. Champagne. "A Theoretically Based Model of Rat Personality with Implications for Welfare." *PLoS ONE* 9 (2014): e95135. doi:10.1371/journal.pone.0095135.

Keim, Brandon. "NIH Decision Signals the Beginning of the End for Medical Research on Chimps." *WIRED*, September 21, 2012. http://www.wired.com/2012/09/nih-chimps-retired/.

Key, Brian. "Why Fish Do Not Feel Pain." *Animal Sentience* (2016): 2016.003.

——. "Falsifying the Null Hypothesis That "Fish Do Not Feel Pain." *Animal Sentience* (2016): 2016.039.

Lents, Nathan H. *Not So Different: Finding Human Nature in Animals*. New York: Columbia University Press, 2016.

Lingle, Susan, and Tobias Riede. "Deer Mothers Are Sensitive to Infant Distress Vocalizations of Diverse Mammalian Species." *The American Naturalist* 184 (2014): 510–22. doi:10.1086/677677.

Panksepp, Jaak, and Jeffrey Burgdorf. "50-kHz Chirping (Laughter?) in Response to Conditioned and Unconditioned Tickle-Induced Reward in Rats: Effects of Social Housing and Genetic Variables." *Behavioural Brain Research* 115 (2000): 25–38. doi:10.1016/S0166-4328(00)00238-2.

"Rats, Mice & Birds." Animal Welfare Institute. https://awionline.org/content/rats-mice-birds.

Roy, Eleanor Ainge. "The Great Escape: Inky the Octopus Legs It to Freedom from Aquarium." *The Guardian*, April 12, 2016. https://www.theguardian.com/world/2016/apr/13/the-great-escape-inky-the-octopus-legs-it-to-freedom-from-new-zealand-aquarium.

Safina, Carl. *Beyond Words: What Animals Think and Feel*. New York: Henry Holt and Co., 2015.

Shettleworth, Sara J. "Clever Animals and Killjoy Explanations in Comparative Psychology." *Trends in Cognitive Science* 14 (2010): 477–81. doi:10.1016/j.tics.2010.07.002.

Striedter, Georg. "Lack of neocortex Does Not Imply Fish Cannot Feel Pain." *Animal Sentience* (2016): 2016.021.

Suddendorf, Thomas. "Mental Time Travel: Continuities and Discontinuities." *Trends in Cognitive Sciences* 17 (2013): 151–52. doi:10.1016/j.tics.2013.01.011.

Suzuki, Toshitaka N. "Communication about Predator Type by a Bird Using Discrete, Graded and Combinatorial Variation in Alarm Calls." *Animal Behaviour* 87 (2014): 59–65. doi:10.1016/j.anbehav.2013.10.009.

Suzuki, Toshitaka, David Wheatcroft, and Michael Griesser. "Experimental Evidence for Compositional Syntax in Bird Calls." *Nature Communications* 7 (2016): 10986. doi:10.1038/ncomms10986.

Urquiza-Haas, Esmeralda G., and Kurt Kotrschal. "The Mind behind Anthropomorphic Thinking: Attribution of Mental States to Other Species." *Animal Behaviour* 109 (2015): 167–76. doi:10.1016/j.anbehav.2015.08.011.

Yoon, Carol Kaesuk. "Donald R. Griffin, 88, Dies; Argued Animals Can Think." *New York Times*, November 14, 2003. http://www.nytimes.com/2003/11/14/nyregion/donald-r-griffin-88-dies-argued-animals-can-think.html.

Honeybees Might Have Emotions

Bateson, Melissa, Suzanne Desire, Sarah E. Gartside, and Geraldine A. Wright. "Agitated Honeybees Exhibit Pessimistic Cognitive Biases." *Current Biology* 21 (2011): 1070–73. doi:10.1016/j.cub.2011.05.017.
Gould, James L. "Honey Bee Cognition." *Cognition* 37 (1990): 83–103. doi:10.1016/0010-0277(90)90019-G.
Gur, Ruben C., et al. "Facial Emotion Discrimination: II. Behavioral Findings in Depression." *Psychiatry Research* 42 (1992): 241–51. doi:10.1016/0165-1781(92)90116-K.
Matheson, Stephanie M., Lucy Asher, and Melissa Bateson. "Larger, Enriched Cages Are Associated with 'Optimistic' Response Biases in Captive European Starlings (*Sturnus vulgaris*)." *Applied Animal Behaviour Science* 109 (2008): 374–83. doi:10.1016/j.applanim.2007.03.007.
Mendl, Michael, Oliver H. P. Burman, Richard M.A. Parker, Elizabeth S. Paul. "Cognitive Bias as an Indicator of Animal Emotion and Welfare: Emerging Evidence and Underlying Mechanisms." *Applied Animal Behaviour Science* 118 (2009): 161–81. doi:10.1016/j.applanim.2009.02.023.
Perry, Clint J., Luigi Baciadonna, and Lars Chittka. "Unexpected Rewards Induce Dopamine-Dependent Positive Emotion–like State Changes in Bumblebees." *Science* 353 (2016): 1529–31. doi: 10.1126/science.aaf4454.

III. Intersections

A Day in the Life of NYC's Hospital for Wild Birds

New Yorkers in Uproar over Planned Mass Killing of Swans

Davis, et al. "Don't Judge Species on Their Origins." *Nature* 474 (2011): 153–54. doi:10.1038/474153a.
"Decline of Submerged Plants in Chesapeake Bay." Chesapeake Bay Field Office, US Fish and Wildlife Service. https://www.fws.gov/chesapeakebay/savpage.htm.
Foderaro, Lisa W. "New York Wants to Banish a Symbol of Love: Mute Swans." *New York Times*, January 29, 2014. http://www.nytimes.com/2014/01/30/nyregion/a-winged-symbol-of-love-that-new-york-state-wants-banished.html?_r=0.
Gayet, et al. "Effects of Mute Swans on Wetlands: A Synthesis." *Hydrobiologia* 723 (2013): 195–204. doi:10.1007/s10750-013-1704-5.
Keim, Brandon. "Ecologists: Time to End Invasive-Species Persecution." *WIRED*, June 8, 2011. http://www.wired.com/2011/06/species-persecution/.
Kraidman, Michelle. "DEC's Final Mute Swan Plan to Come." *Queens Chronicle*, June 23, 2016. http://www.qchron.com/editions/queenswide/dec-s-final-mute-swan-plan-to-come/article_e1367392-f3c1-5649-9a8a-83b0158c2dcc.html.

"New York State Draft Mute Swan Management Plan." Department of Environmental Conservation. http://www.dec.ny.gov/animals/7076.html.

An Eel Swims in the Bronx

Compton, Nette. "Green Infrastructure: Policy and Design." New York City Department of Parks & Recreation. http://www.vitanuova.net/resources/journal/2011_12_02_GI_Webinar.pdf.

Nunn, Mary. "Green Infrastructure in NYC." New York City Department of Parks & Recreation. http://www.houstonlwsforum.org/documents/2013.04.18_GreenInfrastructureinNYC.pdf.

Prosek, James. *Eels: An Exploration, from New Zealand to the Sargasso, of the World's Most Mysterious Fish*. New York: HarperCollins, 2010.

Waldman, John. *Heartbeats in the Muck*. New York: The Lyons Press, 1999.

On Waldman's Pond

Cohen, Mellissa K., and James A. MacDonald. "Northern Snakeheads in New York City." *Northeastern Naturalist* 23 (2016): 11–24. doi 10.1656/045.023.0102.

Coman, Julian. "Scientists Seek Poison to Cull Frankenfish." *The Telegraph*, July 28, 2002. http://www.telegraph.co.uk/news/science/science-news/3297542/Scientists-seek-poison-to-cull-Frankenfish.html.

Odenkirk, John S., and Mike W. Isel. "Trends in Abundance of Northern Snakeheads in Virginia Tributaries of the Potomac River." *Transactions of the American Fisheries Society* 145 (2016): 687–92. doi:10.1080/00028487.2016.1149516.

Waldman, John. *Heartbeats in the Muck*. New York: The Lyons Press, 1999.

The Return of the River

Brown, J. Jed, et al. "Fish and Hydropower on the U.S. Atlantic Coast: Failed Fisheries Policies from Half-Way Technologies." *Conservation Letters* 6 (2013): 280–86. doi:10.1111/conl.12000.

Gardner, C., S. M. Coghlan Jr., J. Zydlewski, and R. Saunders. "Distribution and Abundance of Fishes in Relation to Barriers: Implications for Monitoring Stream Recovery after Barrier Removal." *River Research and Applications* 29 (2013): 65–78. doi:10.1002/rra.1572.

Guyette, Margaret Q., Cynthia S. Loftin, and Joseph Zydlewski. "Carcass Analog Addition Enhances Juvenile Atlantic Salmon (*Salmo salar*) Growth and Condition." *Canadian Journal of Fisheries and Aquatic Sciences* 70 (2013): 860–70. doi:10.1139/cjfas-2012-0496.

Hall, Carolyn J., Adrian Jordaan, and Michael G. Frisk. "The Historic Influence of Dams on Diadromous Fish Habitat with a Focus on River Herring and Hydrologic Longitudinal Connectivity." *Landscape Ecology* 26 (2010): 95–107. doi:10.1007/s10980-010-9539-1.

Hogg, Robert Scott. "Fish Community Response to a Small-Stream Dam Removal in a Maine Coastal River Tributary." MS thesis, University of Maine, 2012.

Holyoke, John. "Pushaw Alewife Restoration a Worthy Plan." *Bangor Daily News*, July 1, 2011. https://bangordailynews.com/2011/07/01/outdoors/holyoke/pushaw-alewife-restoration-a-worthy-plan/.

——. "Edwards Dam Success Foreshadows Penobscot River Project's Future." *Bangor Daily News*, June 8, 2012. https://bangordailynews.com/2012/06/08/outdoors/edwards-dam-success-foreshadows-penobscot-river-projects-future/.

——. "Return of the Shad: Anglers Target Penobscot after 150-Year Hiatus." *Bangor Daily News*, June 16, 2016. https://bangordailynews.com/2016/06/16/outdoors/return-of-the-shad-anglers-target-penobscot-after-150-year-hiatus/.

Leopold, Aldo. *A Sand County Almanac, and Sketches Here and There*. New York: Oxford University Press, 1949.

Miller, Kevin. "Two Years after Dams' Removal, Penobscot River Flourishes." *Portland Press Herald*, September 27, 2015. http://www.pressherald.com/2015/09/27/a-river-revived-the-penobscot-river-two-years-after-dams-removal/.

Mills, Katherine E., Andrew J. Pershing, Timothy F. Sheehan, and David Mountain. "Climate and Ecosystem Linkages Explain Widespread Declines in North American Atlantic Salmon Populations." *Global Change Biology* 19 (2013): 3046–61. doi:10.1111/gcb.12298.

"Penobscot River Restoration: History, Plan and Partners for a Game-Changing Restoration." The Nature Conservancy. http://www.nature.org/ourinitiatives/regions/northamerica/unitedstates/maine/explore/penobscot-river-restoration-project-detail.xml.

"The River." Penobscot River Restoration Trust. http://www.penobscotriver.org/content/4004/the-river.

Trinko Lake, Tara R., Kyle R. Ravana, and Rory Saunders. "Evaluating Changes in Diadromous Species Distributions and Habitat Accessibility following the Penobscot River Restoration Project." *Marine and Coastal Fisheries: Dynamics, Management, and Ecosystem Science* 4 (2012): 284–93. doi:10.1080/19425120.2012.675971.

Royte, J., C. Schmitt and K. Wilson, eds. "Conceptual Monitoring Framework for the Penobscot River Restoration Project." Prepared by the Penobscot River Science Steering Committee (2008). http://www.penobscotriver.org/assets/PRRP_2008_Monitoring_Framework_final.pdf.

Schindler, Daniel E., et al. "Pacific Salmon and the Ecology of Coastal Ecosystems." *Frontiers in Ecology and the Environment* 1 (2003): 31–37. doi:10.1890/1540-9295(2003)001[0031:PSATEO]2.0.CO;2.

Schmitt, Catherine. "Spring Tradition: Anglers Vied to Catch Penobscot's Presidential Salmon." *Bangor Daily News*, March 30, 2012. https://bangordailynews.com/2012/03/30/outdoors/spring-tradition-anglers-vied-to-catch-penobscots-presidential-salmon.

Tucker, Abigail. "On the Elwha, a New Life When the Dam Breaks." Smithsonian.com, September 14, 2011. http://www.smithsonianmag.com/people-places/on-the-elwha-a-new-life-when-the-dam-breaks-79981416/?no-ist=.

Waldman, John. *Running Silver: Restoring Atlantic Rivers and Their Great Fish Migrations*. Guilford, CT: Lyons Press, 2013.

Williams, Ted. "Herring Hearsay." *Fly Rod & Reel*, July/October 2008. http://www.scottchurchdirect.com/docs_ted/herring-hearsay.pdf.

A Chimp's Day in Court: Inside the Historic Demand for Nonhuman Rights

Balter, Michael. "Stone-Throwing Chimp Is Back—And This Time It's Personal." *Science*, May 9, 2012. http://www.sciencemag.org/news/2012/05/stone-throwing-chimp-back-and-time-its-personal.

"Book Discussion on *Rattling the Cage*." C-SPAN, April 4, 2000. http://www.c-span.org/video/?156385-1/book-discussion-rattling-cage.

"Free the Stony Brook Chimpanzees." Stop Animal Exploitation Now! http://saenonline.org/nl-2012-su-07.pdf.

Jensvold, Mary Lee. "Affidavit of Mary Lee Jensvold." http://www.nonhumanrightsproject.org/wp-content/uploads/2013/11/Ex-7-Jensvold-Affidavit-Tommy-Case.pdf.

Keim, Brandon. "Report: Harmful Chimpanzee Research Not Worth the Pain." *WIRED*, December 15, 2011. http://www.wired.com/2011/12/iom-chimp-report/.

——. "Being a Sandpiper." *Aeon*, July 2, 2013. https://aeon.co/essays/what-is-it-like-to-be-a-bird-the-science-of-animal-consciousness.

——. "Leading U.S. Primate Lab Accused of Illegal Chimp Breeding." *WIRED*, November 11, 2014. http://www.wired.com/2011/11/chimpanzee-breeding/.

King, James. "Affidavit of James King." http://www.nonhumanrightsproject.org/wp-content/uploads/2013/11/Ex-8-King-Affidavit-Tommy-Case.pdf.

Larson, Susan G., and Jack T. Stern, Jr. "EMG of Chimpanzee Shoulder Muscles during Knuckle-Walking: Problems of Terrestrial Locomotion in a Suspensory Adapted Primate." *Journal of Zoology* 212 (1987): 629–55. doi:10.1111/j.1469-7998.1987.tb05961.x.

"Latest HSUS Undercover Investigation Reveals Abuse of Chimps, Other Primates in Federally Funded Research Laboratory." The Humane Society of the United States. http://www.humanesociety.org/news/press_releases/2009/03/investigation_chimps_sm_030409.html.

"Legal Documents re. Tommy the Chimpanzee." Nonhuman Rights Project blog, December 2, 2013. http://www.nonhumanrightsproject.org/2013/12/02/legal-documents-re-tommy-kiko-hercules-and-leo-2/.

Matsuzawa, Tetsuro. "Affidavit of Tetsuro Matsuzawa." http://www.nonhumanrightsproject.org/wp-content/uploads/2013/11/Ex-9-Matsuzawa-Affidavit-Tommy-Case.pdf.

McGrew, William. "Affidavit of William C. McGrew." http://www.nonhumanrightsproject.org/wp-content/uploads/2013/12/Suffolk-Ex.-10-McGrew-Affidavit-to-VP.pdf.

Mountain, Michael. "Lawsuit Filed Today on Behalf of Chimpanzee Seeking Legal Personhood." Nonhuman Rights Project blog, December 2, 2013. http://www.nonhumanrightsproject.org/2013/12/02/lawsuit-filed-today-on-behalf-of-chimpanzee-seeking-legal-personhood/.

Osvath, Mathias. "Affidavit of Mathias Osvath." http://www.nonhumanrightsproject.org/wp-content/uploads/2013/11/Ex-11-Osvath-Affidavit-Tommy-Case.pdf.

"Petition for Writ of Habeas Corpus for Tommy." Nonhuman Rights Project, December 2, 2013. http://www.nonhumanrightsproject.org/wp-content/uploads/2013/12/Petition-re-Tommy-Case-Fulton-Cty-NY.pdf.

Posner, Richard A. *How Judges Think*. Cambridge, MA: Harvard University Press, 2010.

Robinson, Marilynne. "'Though the Heavens May Fall' and 'Bury the Chains': Freed." *New York Times*, January 9, 2005. http://www.nytimes.com/2005/01/09/books/review/though-the-heavens-may-fall-and-bury-the-chains-freed.html.

Wise, Steven. *Rattling the Cage: Toward Legal Rights For Animals*. Boston: Da Capo Press, 2000.

——. *Drawing the Line: Science and the Case for Animal Rights*. Cambridge, MA: Perseus Books, 2002.

——. *Though the Heavens May Fall: The Landmark Trial That Led to the End of Human Slavery*. Boston: Da Capo Press, 2005.

Chimpanzee Rights Get a Day in Court

Choplin, Lauren. "Interview with Kevin Schneider re: Tommy, Kiko, Hercules & Leo." Nonhuman Rights Project blog, June 15, 2016. http://www.nonhumanrightsproject.org/2016/06/15/interview-with-kevin-schneider-re-tommy-kiko-hercules-leo/.

"Harvard Law Professor Laurence H. Tribe Submits Amicus Curiae 'Letter-Brief' in Support of The Nonhuman Rights Project." Nonhuman Rights Project blog. http://www.nonhumanrightsproject.org/2015/05/19/harvard-law-professor-laurence-h-tribe-submits-amicus-curiae-letter-brief-in-support-of-the-nonhuman-rights-project/.

Keim, Brandon. "A Chimp's Day in Court: Inside the Historic Demand for Nonhuman Rights." *WIRED*, December 6, 2013. http://www.wired.com/2013/12/chimpanzee-personhood-nonhuman-right/.

——. "Another Court Denies Legal Rights for a Chimpanzee." *WIRED*, January 5, 2015. http://www.wired.com/2015/01/court-denies-kiko-chimp-rights/.

——. "Meet the Chimps That Lawyers Argue Are People." *National Geographic News*, October 8, 2015. http://news.nationalgeographic.com/2015/10/151007-chimps-people-legal-leo-hercules-science/.

Posner, Richard A. "Animal Rights: Legal, Philosophical and Pragmatic Perspectives." In *Animal Rights: Current Debates and New Directions*, edited by Cass R. Sunstein and Martha C. Nussbaum, 51–77. New York: Oxford University Press, 2004.

Walshe, Sabhdh. "Elephants Are People Too (or Soon Could Be)." *Al Jazeera America*, April 13, 2015. http://america.aljazeera.com/articles/2015/4/13/elephants-are-people-too.html.

Yuhas, Alan. "A Third of Americans Believe Animals Deserve Same Rights as People, Poll Finds." *The Guardian*, May 19, 2015. https://www.theguardian.com/world/2015/may/19/americans-animals-human-rights-poll.

Medical Experimentation on Chimps Is Nearing an End. But What about Monkeys?

Brosnan, Sarah F. "Justice- and Fairness-Related Behaviors in Nonhuman Primates." *Proceedings of the National Academy of Sciences* 110, Supplement 2 (2013): 10416–23. doi: Justice- and fairness-related behaviors in nonhuman primates.

Committee on the Use of Chimpanzees in Biomedical and Behavioral Research. *Chimpanzees in Biomedical and Behavioral Research: Assessing the Necessity.* Washington, DC: The National Academies Press, 2011.

Conlee, Kathleen M., and Andrew N. Rowan. "The Case for Phasing Out Experiments on Primates." *Hastings Center Report* 42 (2012): Online supplement. http://animalresearch.thehastingscenter.org/report/the-case-for-phasing-out-experiments-on-primates/.

Ghashghaei H. T., and H. Barbas. "Pathways for Emotion: Interactions of Prefrontal and Anterior Temporal Pathways in the Amygdala of the Rhesus Monkey." *Neuroscience* 115 (2002): 1261–79. doi:10.1016/S0306-4522(02)00446-3.

Keim, Brandon. "Hepatitis C: The Last Chimpanzee Research Battleground." *WIRED*, November 14, 2011. http://www.wired.com/2011/11/chimps-hepatitis-c/.

Melnick, Meredith. "Monkeys, Like Humans, Make Bad Choices and Regret Them, Too." *Time*, May 31, 2011. http://healthland.time.com/2011/05/31/monkeys-play-rock-paper-scissors-and-show-regret-over-losing/.

"Questions and Answers about Monkeys Used in Research." The Humane Society of the United States. http://www.humanesociety.org/animals/monkeys/qa/questions_answers.html.

Rajala, Abigail Z., Katharine R. Reininger, Kimberly M. Lancaster, Luis C. Populin. "Rhesus Monkeys (*Macaca mulatta*) Do Recognize Themselves in the Mirror: Implications for the Evolution of Self-Recognition." *PLoS ONE* 5 (2010): e12865. doi:10.1371/journal.pone.0012865.

Smith, J. David, Joshua S. Redford, Michael J. Beran, David A. Washburn. "Rhesus Monkeys (*Macaca mulatta*) Adaptively Monitor Uncertainty while Multi-tasking." *Animal Cognition* 13 (2010): 93–101. doi:10.1007/s10071-009-0249-5.

Turner, Jonathan. *On the Origins of Human Emotions: A Sociological Inquiry into the Evolution of Human Affect.* Stanford, CA: Stanford University Press, 2000.

I, Cockroach.

Avarguès-Weber, Aurore, and Martin Giurfa. "Conceptual Learning by Miniature Brains." *Proceedings of the Royal Society B* 280 (2013): 20131907. doi:10.1098/rspb.2013.1907.

Bell, William J., Louis M. Roth, and Christine A. Nalepa. *Cockroaches: Ecology, Behavior, and Natural History*. Baltimore: Johns Hopkins University Press, 2007.

Campbell, Robert A. A., and Glenn C. Turner. "The Mushroom Body." *Current Biology* 20 (2010): R11–R12. doi:10.1016/j.cub.2009.10.031.

Cheeseman, James F., et al. "Way-Finding in Displaced Clock-Shifted Bees Proves the Bee Reads Its Map." *Proceedings of the National Academy of Sciences*, 111 (2014): 8949–54. doi:10.1073/pnas.1408039111.

Crist, Eileen. "The Inner Life of Earthworms: Darwin's Argument and Its Applications." In *The Cognitive Animal: Empirical and Theoretical Perspectives on Animal Cognition*, edited by Marc Bekoff, Colin Allen, and Gordon M. Burghardt, 3–8. Cambridge, MA: The MIT Press, 2002.

Crook, Robyn Jean. "The Welfare of Invertebrate Animals in Research: Can Science's Next Generation Improve Their Lot?" *Journal of Postdoctoral Research* 1 (2013): 9–20.

"Ethical Issues regarding the Use of Invertebrates in Education." Backyard Brains. http://wiki.backyardbrains.com/Ethical_Issues_Regarding_the_Use_of_Invertebrates_in_Education.

Griffin, Donald, and Gayle B. Speck. "New Evidence of Animal Consciousness." *Animal Cognition* 7 (2004): 5–18. doi:10.1007/s10071-003-0203-x.

Koch, Christof. "Ubiquitous Minds." *Scientific American Mind* 25 (2014): 26–29. doi:10.1038/scientificamericanmind0114-26.

Lihoreau, M., J. T. Costa, and C. Rivault. "The Social Biology of Domiciliary Cockroaches: Colony Structure, Kin Recognition and Collective Decisions." *Insectes Sociaux* 59 (2012): 445–52. doi:10.1007/s00040-012-0234-x.

Menzel, Randolf. "Insect Minds for Human Minds." In *Human Learning: Biology, Brain and Neurocience*, edited by Mark Guadagnoli, 271. Amsterdam: North-Holland, 2008.

——. "The Honeybee as a Model for Understanding the Basis of Cognition." *Nature Reviews Neuroscience* 13 (2012): 758–68. doi: 10.1038/nrn3357.

Menzel, Randolf, and Bjorn Brembs. "Cognition in Invertebrates." In *Evolution of Nervous Systems*, vol. 1, *Theories, Development, Invertebrates*, edited by J. H. Kaas. Oxford: Academic Press, 2007.

Menzel, Randolf, and Martin Giurfa. "Dimensions of Cognition in an Insect, the Honeybee." *Behavioral and Cognitive Neuroscience Reviews* 5 (2006): 24–40. doi:10.1177/1534582306289522.

"The RoboRoach: Control a Living Insect from your Smartphone!" Kickstarter. https://www.kickstarter.com/projects/backyardbrains/the-roboroach-control-a-living-insect-from-your-sm.

Sakura, Midori, and Makoto Mizunami. "Olfactory Learning and Memory in the Cockroach *Periplaneta americana*." *Zoological Science* 18 (2001): 21–28. doi:10.2108/zsj.18.21.

Tononi, Giulio. "Integrated Information Theory of Consciousness: An Updated Account." *Archives Italiennes de Biologie* 150 (2012): 290–326. doi:10.4449/aib.v149i5.1388.

IV. Ethics

The Improbable Bee

"Autonomous Flying Microrobots (RoboBees)." Wyss Institute. http://wyss.harvard.edu/viewpage/457.

Cummins, Kate. "UK Engineers Develop Bee-Based Security Device." *The Engineer*, March 28, 2011. https://www.theengineer.co.uk/issues/28-march-2011/uk-engineers-develop-bee-based-security-device/.

Keim, Brandon. "Beyond Black and Yellow: The Stunning Colors of America's Native Bees." *WIRED*, August 12, 2013. http://www.wired.com/2013/08/beautiful-bees/.

Spector, Dina. "Tiny Flying Robots Are Being Built to Pollinate Crops Instead of Real Bees." *Business Insider*, July 7, 2014. http://www.businessinsider.com/harvard-robobees-closer-to-pollinating-crops-2014-6.

Wired Science. "Tiny New Compound Camera Is Built like a Bug's Eye." *WIRED*, May 1, 2013. http://www.wired.com/2013/05/bugs-eye-camera/.

The Ethics of Urban Beekeeping

Beurteaux, Danielle. "Now Cleared for Landing at Airports: Bees." *New York Times*, February 19, 2015. http://www.nytimes.com/2015/02/24/science/now-cleared-for-landing-at-airports-bees.html?ref=topics.

Keim, Brandon. "Beyond Black and Yellow: The Stunning Colors of America's Native Bees." *WIRED*, August 12, 2013. http://www.wired.com/2013/08/beautiful-bees/.

——. "Beyond Honeybees: Now Wild Bees and Butterflies May Be in Trouble." *WIRED*, May 5, 2014. http://www.wired.com/2014/05/wild-bee-and-butterfly-declines/.

——. "Honeybees vs. Native Bees: Literature Excerpts." Whalefall blog. http://whalefall.tumblr.com/post/117018572880/honeybees-vs-native-bees-literature-excerpts.

Matteson, Kevin C. "A Pictorial Guide to Some Common Bees of the New York City Metropolitan Area." The Great Pollinator Project. http://greatpollinatorproject.org/sites/all/downloads/pdfs/Pictorial_Guide_to_Common_NYC_Bees.pdf.

Matteson, Kevin C., and Gail A. Langellotto. "Determinates of Inner City Butterfly and Bee Species Richness." *Urban Ecosystems* 13 (2010): 333–47. doi:10.1007/s11252-010-0122-y.

Schmetterling, David. “As Pollinators for a Native Plant Garden, Honey Bees Suck!” *Garden Rant*, March 8, 2012. http://gardenrant.com/2012/03/as-pollinators-for-a-native-plant-garden-honey-bees-suck.html.

Vorpal, Kaz. “List of Crops That Don’t Need Honeybees.” Facebook post, July 9, 2013. https://www.facebook.com/notes/kaz-vorpal/list-of-crops-that-dont-need-honeybees/10151447302312854.

The Wild, Secret Life of New York City

Crist, Eileen. “Ptolemaic Environmentalism.” In *Keeping the Wild: Against the Domestication of Earth*, edited by George Wuerthner, Eileen Crist, and Tom Butler, 16–30. Washington, DC: Island Press, 2014.

Del Tredici, Peter. “Spontaneous Urban Vegetation: Reflections of Change in a Globalized World.” *Nature and Culture* 5 (2010): 299–315. doi:10.3167/nc.2010.050305.

——. *Wild Urban Plants of the Northeast: A Field Guide*. Ithaca, NY: Cornell University Press, 2010.

——. “The Flora of the Future.” *Places*, April 2014. https://placesjournal.org/article/the-flora-of-the-future/.

“High Line Plant List.” http://assets.thehighline.org/pdf/12_High%20Line%20Plant%20List.pdf.

Kaye, Roger. “What Future for the Wildness of Wilderness in the Anthropocene?” *Alaska Park Science* 13 (2014): 41–45.

Kowarik, Ingo. “Novel Urban Ecosystems, Biodiversity, and Conservation.” *Environmental Pollution* 159 (2011): 1974–83. doi:10.1016/j.envpol.2011.02.022.

Kowarik, Ingo, and Andreas Langer. “Natur-Park Südgelände: Linking Conservation and Recreation in an Abandoned Railyard in Berlin.” In *Wild Urban Woodlands*, edited by Ingo Kowarik and Stefan Körner, 287–99. Springer-Verlag Berlin Heidelberg, 2005.

Maurer, U., T. Peschel, and S. Schmitz. “The Flora of Selected Urban Land-Use Types in Berlin and Potsdam with regard to Nature Conservation in Cities.” *Landscape and Urban Planning* 46 (2000): 209–15. doi:10.1016/S0169-2046(99)00066-3.

McKinney, Michael L. “Urbanization as a Major Cause of Biotic Homogenization.” *Biological Conservation* 127 (2006): 247–60. doi:10.1016/j.biocon.2005.09.005.

Nash, Roderick. “Power of the Wild.” *New Scientist*, March 30, 2002. https://www.newscientist.com/article/mg17323364-700-power-of-the-wild/.

Pyšek, Petr. et al. “Trends in Species Diversity and Composition of Urban Vegetation over Three Decades.” *Journal of Vegetation Science* 15 (2004): 781–88. doi:10.1111/j.1654-1103.2004.tb02321.x.

Sagoff, Mark. “Who Is the Invader? Alien Species, Property Rights, and the Police Power.” *Social Philosophy and Policy* 26 (2009): 26–52. doi:10.1017/S0265052509090165.

Stalter, Richard. "The Flora on the High Line, New York City, New York." *The Journal of the Torrey Botanical Society* 131 (2004): 387–93. doi:10.2307/4126942.

Vessel, Matthew F., and Herbert H. Wong. *Natural History of Vacant Lots.* Berkeley: University of California Press, 1987.

Earth Is Not a Garden

Aronson, et al. "A Global Analysis of the Impacts of Urbanization on Bird and Plant Diversity Reveals Key Anthropogenic Drivers." *Proceedings of the Royal Society B* 281 (2014): 20133330. doi:10.1098/rspb.2013.3330.

Barlow, Jos., et al. "How Pristine Are Tropical Forests? An Ecological Perspective on the Pre-Columbian Human Footprint in Amazonia and Implications for Contemporary Conservation." *Biological Conservation* 151 (2012): 45–49. doi:10.1016/j.biocon.2011.10.013.

"Bringing Balance to Colombia's Magdalena River." The Nature Conservancy. http://www.nature.org/ourinitiatives/regions/southamerica/colombia/explore/bringing-balance-to-colombias-magdalena-river.xml.

Bush, Mark B., and Miles R. Silman. "Amazonian Exploitation Revisited: Ecological Asymmetry and the Policy Pendulum." *Frontiers in Ecology and the Environment* 5 (2007): 457–65. doi:10.1890/070018.

Crist, Eileen. "Ptolemaic Environmentalism." In *Keeping the Wild: Against the Domestication of Earth*, edited by George Wuerthner, Eileen Crist, and Tom Butler, 16–30. Washington, DC: Island Press, 2014.

Frank, Adam. "A Human-Driven Mass Extinction: Good or Bad?" *NPR*, January 28, 2014. http://www.npr.org/sections/13.7/2014/01/28/267038785/a-human-driven-mass-extinction-good-or-bad.

Heckenberger, et al. "Amazonia 1492: Pristine Forest or Cultural Parkland?" *Science* 301 (2003): 1710–14. doi:10.1126/science.1086112.

Johns, David. "With Friends Like These, Wilderness and Biodiversity Do Not Need Enemies." In Wuerthner, Crist, and Butler, *Keeping the Wild*, 31-44.

Kareiva, Peter, Sean Watts, Robert McDonald, and Tim Boucher. "Domesticated Nature: Shaping Landscapes and Ecosystems for Human Welfare." *Science* 316 (2007): 1866–69. doi:10.1126/science.1140170.

——. "Conservation in the Anthropocene: Beyond Solitude and Fragility." *Breakthrough Journal*, Winter 2012. http://thebreakthrough.org/index.php/journal/past-issues/issue-2/conservation-in-the-anthropocene.

Kaye, Roger. "What Future for the Wildness of Wilderness in the Anthropocene?" *Alaska Park Science* 13 (2014): 41–45.

Keim, Brandon. "Return of the Ghost Fish." *On Earth*, November 12, 2013. http://archive.onearth.org/articles/2013/11/can-civilization-and-salmon-coexist-dam-good-question.

Kloor, Keith. "The Green Modernist Vision." Collide-a-Scape blog. http://blogs.discovermagazine.com/collideascape/2012/04/17/the-green-modernist-vision/.

Latour, Bruno. "Love Your Monsters: Why We Must Care for Our Technologies as We Do Our Children." *Breakthrough Journal*, Winter 2012. http://thebreakthrough.org/index.php/journal/past-issues/issue-2/love-your-monsters.

Leopold, Aldo. *A Sand County Almanac*. New York: Oxford University Press, 1949.

Levikov, Nika. "Life Finds a Way: Surprising Biodiversity in Cities." *Mongabay*, April 11, 2014. https://news.mongabay.com/2014/04/life-finds-a-way-the-surprising-biodiversity-of-cities/.

Marris, Emma. *The Rambunctious Garden: Saving Nature in a Post-Wild World*. New York: Bloomsbury USA, 2011.

Marris, Emma, Peter Kareiva, Joseph Mascaro, and Erle C. Ellis. "Hope in the Age of Man." *New York Times*, December 7, 2011. http://www.nytimes.com/2011/12/08/opinion/the-age-of-man-is-not-a-disaster.html.

Marshall, Robert. "The Problem of the Wilderness." *The Scientific Monthly* 30 (1930): 141–48.

Nash, Roderick. *Wilderness and the American Mind*. New Haven, CT: Yale University Press, 1967.

——. "Why Wilderness?" *Earth Island Journal*, September 3, 2014. http://www.earthisland.org/journal/index.php/elist/eListRead/why_wilderness/.

Nordhaus, Ted, and Michael Shellenberger. "Evolve: The Case for Modernization as the Road to Salvation." *Breakthrough Journal*, Winter 2012. http://thebreakthrough.org/index.php/journal/past-issues/issue-2/evolve.

Shellenberger, Michael, and Ted Nordhaus, eds. *Love Your Monsters: Postenvironmentalism and the Anthropocene*. Oakland, CA: Breakthrough Institute, 2011. Kindle edition.

Snyder, Gary. *The Practice of the Wild*. San Francisco: North Point Press, 1990.

——. *A Place in Space: Ethics, Aesthetics, and Watersheds*. Berkeley: Counterpoint Press, 1995.

Thoreau, Henry David. *Winter: From the Journal of Henry D. Thoreau*, edited by Harrison Gray Otis Blake. Concord, MA: Houghton, Mifflin, 1891.

Waldman, John. *Running Silver: Restoring Atlantic Rivers and Their Great Fish Migrations*. Guilford, CT: Lyons Press, 2013.

Add a Few Species. Pull Down the Fences. Step Back.

Kareiva, Peter, Michelle Marvier, and Robert Lalasz. "Conservation in the Anthropocene: Beyond Solitude and Fragility." *Breakthrough Journal*, Winter 2012. http://thebreakthrough.org/index.php/journal/past-issues/issue-2/conservation-in-the-anthropocene.

Marris, Emma. *The Rambunctious Garden: Saving Nature in a Post-Wild World*. New York: Bloomsbury USA, 2011.

Monbiot, George. *Feral: Rewilding the Land, the Sea, and Human Life*. Chicago, IL: The University of Chicago Press, 2014.

Feral Cats vs. Conservation: A Truce

"1080 Poison." The World League of Protection for Animals. http://www.wlpa.org/1080_poison.htm.

Brook, Leila A., Christopher N. Johnson, and Euan G. Ritchie. "Effects of Predator Control on Behaviour of an Apex Predator and Indirect Consequences for Mesopredator Suppression." *Journal of Applied Ecology* 49 (2012): 1278–86.

Condon, Michael. "Feral Cat Control Needs a National Approach, Says Threatened Species Commissioner." *ABC Rural*, April 13, 2015. http://www.abc.net.au/news/2015-04-13/feral-cat-control-needs-a-national-approach/6388324.

Grubbs, Shannon E., and Paul R. Krausman. "Observations of Coyote-Cat Interactions." *The Journal of Wildlife Management* 73 (2009): 683–85. doi:10.2193/2008-033.

von Hörchner, Cherie. "'Let Dingoes Eat Kangaroos' Says University of Queensland Study." *ABC Rural*, January 16, 2015. http://www.abc.net.au/news/2015-01-16/dingo-roo-cattle-study/6021808.

Wallach, Arian D., William J. Ripple, and Scott P. Carroll. "Novel Trophic Cascades: Apex Predators Enable Coexistence." *Trends in Ecology and Evolution* 30 (2015): 146–53. doi:10.1016/j.tree.2015.01.003.

Woinarski, John C. Z., Andrew A. Burbidge, and Peter L. Harrison. "Ongoing Unraveling of a Continental Fauna: Decline and Extinction of Australian Mammals since European Settlement." *Proceedings of the National Academy of Sciences* 112 (2015): 4531–40. doi:10.1073/pnas.1417301112.

Should Animals Have a Right to Privacy?

Brown, JPat. "New Jersey Rejects Request for Dolphin Necropsy Results, Citing 'Medical Privacy.'" MuckRock, January 11, 2016. https://www.muckrock.com/news/archives/2016/jan/11/new-jersey-FOIA-dolphin-medical-privacy/.

Domonoske, Camila. "Monkey Can't Own Copyright to His Selfie, Federal Judge Says." *NPR*, January 7, 2016. http://www.npr.org/sections/thetwo-way/2016/01/07/462245189/federal-judge-says-monkey-cant-own-copyright-to-his-selfie.

Jewell, Zoe. "Effect of Monitoring Technique on Quality of Conservation Science." *Conservation Biology* 27 (2013): 501–8. doi:10.1111/cobi.12066.

Koebler, Jason. "New Jersey Says Releasing Dolphin's Autopsy Would Infringe Its Privacy." *Motherboard*, January 11, 2016. https://motherboard.vice.com/read/new-jersey-says-releasing-dead-dolphins-autopsy-would-infringe-its-privacy.

Mortillaro, Nicole. "Is This Polar Bear Really Being Choked by a Research Collar?" *Global News*, November 23, 2015. http://globalnews.ca/news/2331033/is-this-polar-bear-really-being-choked-by-a-research-collar/.

"NJ Dolphin Necropsy." MuckRock. https://www.muckrock.com/foi/new-jersey-229/nj-dolphin-necropsy-20430/#comm-218363.

When Climate Change Blinds Us

Greenfieldboyce, Nell. "Why Is It So Hard to Save Gulf of Maine Cod? They're in Hot Water." *NPR*, October 29, 2015. http://www.npr.org/sections/thesalt/2015/10/29/451942641/why-is-it-so-hard-to-save-gulf-of-maine-cod-theyre-in-hot-water.

Jackson, Derrick Z. "Climate Change Fuels Cod Collapse." *Boston Globe*, November 3, 2015. https://www.bostonglobe.com/opinion/2015/11/03/climate-change-fuels-cod-collapse/lyCrsy0atWxs2sNSFccOFL/story.html.

Pershing, Andrew J., et al. "Slow Adaptation in the Face of Rapid Warming Leads to Collapse of the Gulf of Maine Cod Fishery." *Science* 350 (2015): 809–12. doi:10.1126/science.aac9819.

Pope Francis. *Laudato Si: On Care for Our Common Home*, 2015. http://w2.vatican.va/content/francesco/en/encyclicals/documents/papa-francesco_20150524_enciclica-laudato-si.html.

Terry, Rebecca C., and Rebecca J. Rowe. "Energy Flow and Functional Compensation in Great Basin Small Mammals under Natural and Anthropogenic Environmental Change." *Proceedings of the National Academy of Sciences* 112 (2015): 9656–61. doi:10.1073/pnas.1424315112.

To Bring Back Extinct Species, We'll Need to Change Our Own

Brand, Stewart. "The Dawn of De-extinction. Are You Ready?" TED.com, March 2013. https://www.ted.com/talks/stewart_brand_the_dawn_of_de_extinction_are_you_ready/transcript?language=en#t-529689.

Dell'Amore, Christine. "Death of Rare White Rhino Leaves 5 in the World." *National Geographic News*, December 17, 2014. http://news.nationalgeographic.com/news/2014/12/141216-rhinoceros-death-breeding-science-world-endangered-animals/.

"Passenger Pigeons in Your State, Province or Territory: New York." Project Passenger Pigeon. http://passengerpigeon.org/states/NewYork.html.

Switek, Brian. "The Promise and Pitfalls of Resurrection Ecology." Laelaps blog, March 12, 2013. http://phenomena.nationalgeographic.com/2013/03/12/the-promise-and-pitfalls-of-resurrection-ecology/.

——. "Reinventing the Mammoth." *National Geographic News*, March 19, 2013. http://phenomena.nationalgeographic.com/2013/03/19/reinventing-the-mammoth/.

"TEDxDeExtinction." Revive & Restore. https://www.ted.com/tedx/events/7650.

Zimmer, Carl. "Bringing Extinct Species Back to Life." *National Geographic*, April 2013. http://ngm.nationalgeographic.com/2013/04/125-species-revival/zimmer-text.

September 11, Fall Migration, and Occupy Wall Street

Friedman, Jeffrey, and Peter J. Wallison. "A Crisis of Politics, Not Economics: Complexity, Ignorance, and Policy Failure." *Critical Review* 20 (2009): 127–83. doi:10.1080/08913810903030980.

Keim, Brandon. "9/11 Memorial Lights Trap Thousands of Birds." *WIRED*, September 14, 2010. http://www.wired.com/2010/09/tribute-in-light-birds/.
Stiglitz, Joseph E. "The Anatomy of a Murder: Who Killed America's Economy?" *Critical Review* 20 (2009): 329–99. doi:10.1080/08913810902934133.
White, Lawrence J. "The Credit-Rating Agencies and the Subprime Debacle." *Critical Review* 20 (2009): 389–99. doi:10.1080/08913810902974964.

Making Sense of 7 Billion People

Barnosky, Anthony D., et al. "Has the Earth's Sixth Mass Extinction Already Arrived?" *Nature* 471 (2011): 51–57. doi:10.1038/nature09678.
Erb, Karl-Heinz, et al. "Analyzing the Global Human Appropriation of Net Primary Production—Processes, Trajectories, Implications: An Introduction." *Ecological Economics* 69 (2009): 250–59. doi:10.1016/j.ecolecon.2009.07.001.
Estes, James A., et al. "Trophic Downgrading of Planet Earth." *Science* 333 (2011): 301–6. doi:10.1126/science.1205106.
Forbes, Peter. "The God Species by Mark Lynas—Review." *The Guardian*, July 20, 2011. https://www.theguardian.com/books/2011/jul/20/mark-lynas-god-species-review.
Rettie, James C. "But a Watch in the Night." http://socialmachines.media.mit.edu/wp-content/uploads/sites/27/2015/02/Rettie-But-a-Watch-in-the-Night.pdf.
Sanderson, Eric W., et al. "The Human Footprint and the Last of the Wild." *BioScience* 52 (2002): 891–904. doi:10.1641/0006-3568(2002)052[0891:THFATL]2.0.CO;2.
Whiteside, et al. "Delayed Recovery of Non-Marine Tetrapods after the End-Permian Mass Extinction Tracks Global Carbon Cycle." *Proceedings of the Royal Society B* 279 (2012): 1310–19. doi:10.1098/rspb.2011.1895.
Wired Science. "World Population: A National Geographic Tour." *WIRED*, October 31, 2011. http://www.wired.com/2011/10/7_billion_gallery/.

All URLs in the references section were accessed July 18, 2016.

INDEX